Springer Proceedings in Complexity

More information about this series at
http://www.springer.com/series/11637

Springer Complexity

Springer Complexity is an interdisciplinary program publishing the best research and academic-level teaching on both fundamental and applied aspects of complex systems — cutting across all traditional disciplines of the natural and life sciences, engineering, economics, medicine, neuroscience, social and computer science.

Complex Systems are systems that comprise many interacting parts with the ability to generate a new quality of macroscopic collective behavior the manifestations of which are the spontaneous formation of distinctive temporal, spatial or functional structures. Models of such systems can be successfully mapped onto quite diverse "real-life" situations like the climate, the coherent emission of light from lasers, chemical reaction-diffusion systems, biological cellular networks, the dynamics of stock markets and of the internet, earthquake statistics and prediction, freeway traffic, the human brain, or the formation of opinions in social systems, to name just some of the popular applications.

Although their scope and methodologies overlap somewhat, one can distinguish the following main concepts and tools: self-organization, nonlinear dynamics, synergetics, turbulence, dynamical systems, catastrophes, instabilities, stochastic processes, chaos, graphs and networks, cellular automata, adaptive systems, genetic algorithms and computational intelligence.

The three major book publication platforms of the Springer Complexity program are the monograph series "Understanding Complex Systems" focusing on the various applications of complexity, the "Springer Series in Synergetics", which is devoted to the quantitative theoretical and methodological foundations, and the "SpringerBriefs in Complexity" which are concise and topical working reports, case-studies, surveys, essays and lecture notes of relevance to the field. In addition to the books in these two core series, the program also incorporates individual titles ranging from textbooks to major reference works.

Editorial and Programme Advisory Board

Katharina Zweig • Wolfgang Neuser •
Volkmar Pipek • Markus Rohde • Ingo Scholtes
Editors

Socioinformatics - The Social Impact of Interactions between Humans and IT

 Springer

Editors

Katharina Zweig
Department of Computer Science
TU Kaiserslautern
Kaiserslautern, Germany

Volkmar Pipek
Institut für Wirtschaftsinformatik
Universität Siegen
Siegen, Germany
and
International Institute for Socio-Informatics
   (IISI), Bonn
Germany

Wolfgang Neuser
Faculty Social Sciences
TU Kaiserslautern
Kaiserslautern, Germany

Markus Rohde
Institut für Wirtschaftsinformatik
Universität Siegen
Siegen, Germany
and
International Institute for Socio-Informatics
   (IISI), Bonn
Germany

Ingo Scholtes
Chair of Systems Design
ETH Zürich
Zürich, Switzerland

ISSN 2213-8684   ISSN 2213-8692 (electronic)
ISBN 978-3-319-09377-2   ISBN 978-3-319-09378-9 (eBook)
DOI 10.1007/978-3-319-09378-9
Springer Cham Heidelberg New York Dordrecht London

Library of Congress Control Number: 2014949770

Printed on acid-free paper

Springer is part of Springer Science+Business Media (www.springer.com)

Contents

Introduction

Katharina Anna Zweig, Wolfgang Neuser, Ingo Scholtes, Volkmar Pipek, and Markus Rohde

Abstract Software has in many cases become a part of a socio-technical system that needs new modeling approaches to understand the interaction between IT-systems, the individual, organizations, and the society at large.

Socioinformatics

This book combines the entries to our workshop on Socioinformatics at the 43rd annual meeting of the German Informatics Society (Gesellschaft für Informatik) in Koblenz.

The term *socioinformatics* is rather new and may thus need some explanation: In the early days of computer science, software was mainly a simple product which made a well-known process faster and less error-prone. For example, without word processing software, all texts had to be written by hand or typed on a typewriter; every mistake required to start all over again. Software enabled the typist to simply erase the wrong letters, leaving the rest of the document intact. Today, collecting

K.A. Zweig (✉)
Department of Computer Science, Graph Theory and Complex Network Analysis, University Kaiserslautern, Gottlieb-Daimler-Strasse 49, 67663 Kaiserslautern, Germany
e-mail: zweig@cs.uni-kl.de

W. Neuser
Faculty Social Sciences, TU Kaiserslautern, Kaiserslautern, Germany
e-mail: neuser@rhrk.uni-kl.de

I. Scholtes
Chair of Systems Design, ETH Zürich, Weinbergstrasse 56/58, 8092 Zürich, Switzerland
e-mail: ischoltes@ethz.ch

V. Pipek • M. Rohde
Institut für Wirtschaftsinformatik, Universität Siegen, Hölderlin 3, 57076 Siegen, Germany

International Institute for Socio-Informatics (IISI), Dorotheenstr. 76, 53111 Bonn, Germany
e-mail: volkmar.pipek@unisiegen.de; markus.rohde@iisi.de

data and processing it digitally is so fast that processes emerge that have no equivalent to the processes of the pre-computer era, for example the decentralized organization of a power-grid fueled by renewable energy sources. IT-systems thus have an unprecedented influence on us as individuals, on the organizations we are embedded in, and on society at large—together we build a so-called socio-technical system; these systems have to be carefully designed to make them acceptable for humans. Jörg Dörr's entry with the title "Towards Acceptance of Socio-Technical Systems: An Emphasis on the Requirements Phase" (Chap. 10) focuses on how requirements engineering can help to build better systems. But designing software for these socio-technical systems also requires new models, as proposed in the entry "Integrated Modeling and Evolution of Social Software" (Chap. 6) by Arnd Poetzsch-Heffter, Barbara Paech, and Mathias Weber. The interaction between IT-systems and humans is largely determined by human behavior—thus, a socioinformatician needs to understand human constraints. Janet Siegmund and Sven Apel's entry on "The Human Factor in Computer Science and How to Teach Students to Care: An Experience Report" (Chap. 2) summarizes their experience in teaching a course on empirical studies in software design. These first three entries are mainly concerned with software engineering aspects of creating socio technical systems.

The following entries focus on technical systems that support social processes. The entry by Kai Fischbach, Oliver Posegga, and Martin Donath with the title "Using Weighted Interaction Metrics for Link Prediction in a Large Online Social Network" (Chap. 5) shows that knowing the frequencies by which users of an online social network interact helps to predict new links. These methods can be used to introduce persons to each other to facilitate new friendships, thus supporting a social process which would normally be mediated by a common friend. Jean Botev's entry with the title "Anonymity, Immediacy and Electoral Delegation in Socio-Technical Computer Systems" (Chap. 9) gives an overview on software that supports collective decision making. Finally, Michael Seufert, George Darzanos, Ioanna Papafili, Roman Łapacz, Valentin Burger, and Tobias Hoßfeld discuss "Socially-Aware Traffic Management" (Chap. 3) which uses social knowledge to create better document distributions in peer-to-peer networks.

The last section of the book contains five entries that are focusing on the co-evolution of society and software and on the possible mutual influence of IT-systems and organizations. Ingo Scholtes, René Pfitzner, and Frank Schweitzer in their entry "The Social Dimension of Information Ranking: A Discussion of Research Challenges and Approaches" (Chap. 4) discuss how social processes (like tagging and rating information) help to rank large amounts of documents but also how these processes create new incentives and might change social processes in turn. In his entry with the title "Towards a Principle of Socio-Technical Interactions: Embracing Cultural Issues of Enterprise Culture Through a Concept of Enterprise Activities" (Chap. 1), Sebastian Bittman models the interaction of human and mechanical agents, based on Max Weber's theory of social action. Valentin Burger et al. in their article on "Social Network Analysis in the Enterprise: Challenges and Opportunities (Chap. 7)" focus on the question of how enterprise software changes the way workers can be evaluated, while Sean P. Goggins and Giuseppe Valetto in their chapter on "Assessing the Structural Fluidity of Virtual Organizations and Its Effects" (Chap. 8) elucidate how IT-systems support the building and evolution of virtual organizations, where organizations can cooperate in a highly dynamic manner. Finally, Wolfgang Lenski, in his entry "Morals, IT-Structures, and Society" (Chap. 11), sketches a philosophical framework describing how morality, society, and IT-structures mutually influence each other.

The workshop and this proceedings has shown that there are many aspects of today's IT-systems which intertwine them heavily with how we, as humans, live our daily lifes. The organizers of the workshop hope that the workshop will become a regular meeting place for all computer scientists, sociologists, psychologists, lawyers, and philosophers who want to understand this field at the intersection of all of the aforementioned disciplines, a field called *Socioinformatics*.

Chapter 1
Towards a Principle of Socio-technical Interactions – Embracing Cultural Issues of Enterprise Culture Through a Concept of Enterprise Activities

Sebastian Bittmann

Abstract Due to the increasing importance of Information Technology, the execution of actions in the context of enterprises requires more than in the last decades, the alignment between concerned actors and additionally the cooperation between them, in particular human and mechanical agents. However, through the progressing automation and autonomisation of information technology, the cooperation with mechanical agents is still necessary. To embrace specific consequences of such developments, a theory of action in enterprise environments will be developed on the basis of MAX WEBER'S theory of social action, which explains enterprise actions of agents in an enterprise. Enterprise actions are not only oriented towards the interaction with other agents in the social system, but consider additionally the requirements that were formulated for the agent by the enterprise management. Despite these requirements, agents are able to contribute to the enterprise culture by varying and adapting their actions. In this way, every agent contributes not only to the economic, but also to the social and cultural capital. By introducing IT systems as autonomously acting agents, the balance between economic, social and cultural capital is endangered. Any kind of creativity will be prohibited by unambiguous instructions and the execution of fully automated actions is completely oriented towards to enterprise strategy, without considering important social and cultural factors. Thereby, this paper introduces and discusses a general conception of socio-technological interactions within enterprises. Specifically, the concept of such enterprise actions will be used to illustrate how enterprise culture evolves and how it is threatened by strict and standardised socio-technological interactions.

S. Bittmann (✉)
Information Management and Information Systems, University Osnabrueck,
Osnabrück, Germany
e-mail: sebastian.bittmann@uni-osnabrueck.de

© Springer International Publishing Switzerland 2014   1
K. Zweig et al. (eds.), *Socioinformatics - The Social Impact of Interactions between Humans and IT*, Springer Proceedings in Complexity,
DOI 10.1007/978-3-319-09378-9_1

1.1 Introduction

Business process specifications generally describe the organisational behaviour of an enterprise [7, 12, 25]. In particular, this scheme gained importance in interdisciplinary collaboration as a research field as well as practical applications, through emergency of alignment possibilities, especially between the business perspective and the required development of IT to support business processes [5, 24]. Recently, this development is promoted by the necessary alignment of activities, which cannot be aligned through collaborative work. Through the usage of information technology to execute business processes, two kinds of performing agents can be identified. In addition to the classical agent, which is the human individual and the employee, coexists the mechanical agent, manifested in program code. This agent deserves the label 'agent' today even more than in the previous decades, since it is efficiently able to identify work orders autonomously and to react to them. Through the increasing capability of IT systems to adapt to their environment and the growing digitalization, it is now rarely required of human agents to specify work orders and thus to conform the orders to currently existing specifics, given that IT systems can autonomously estimate the effort and process the orders [22]. By means of the progressing development of information systems it is possible to specify rules for machines to autonomously carry out their work without any human cooperation [18]. Nonetheless machines are potentially dependent on preliminary work of human agents, as well as human agents are potentially dependent on preliminary work of machines. Thus there are two different types of agents in an enterprise, which cooperate according to the paradigm of the business process [7] to achieve the enterprise objective. But both of them can be completely autonomous in their work and none of them is in any way instrument for the other one. Figure 1.1 illustrates the emergence of the mechanical agent within the enterprise through the exclusion of the human agent from task accomplishment. The IT evolves from a mere instrument to an autonomous agent, not any longer used for a specific task accomplishment, but now independently responsible for own tasks.

The interaction between the human and mechanical agents has to be investigated with respect to social and cultural phenomena, such as the enterprise culture [11]. Hereby is in question the formation of such enterprise culture without the opportunities of creative freedom, which enables an increase in cultural and social capital with the help of unplanned actions [15]. The latter of the paper is structured as follows. Initially the theory of social action by MAX WEBER will be introduced in Sect. 1.2. Section 1.3 presents an extension of Weber's theory with a specialization of social action into a theory of enterprise action. Following this, it is illustrated which factors influence these enterprise actions. Later on the enterprise actions are analysed with regard to the two types of agents introduced in the previous paragraph. This analysis will be conducted using the previously discussed business process models to show the specific roles of the agents. The paper ends with a conclusion.

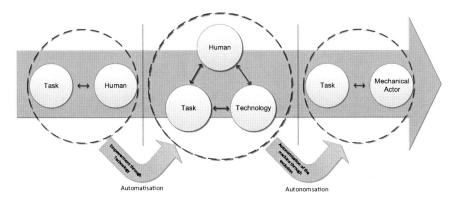

Fig. 1.1 Emergence of the automatic actor, with reference to ([13]: p. 16)

1.2   Max Weber's Theory of Social Action

WEBER defines the term of social action as a special form of human behaviour. Behaviour is every kind of human activity and can be covert, i.e. not observable by others, as well as overt. Behaviour comprises, in addition, both omission and acquiescence. An action, according to WEBER, is a special form of behaviour, which is associated with a sense. Since this sense is subjective, the actions do not have to adhere to social norms or other conditions. In fact the attributed meaning does not have to be meaningful for anyone else than the actor itself. WEBER motivates the dependency by the experience of an action on the understanding of the meaning of that particular action, but not on the interpretation of such meaning. Due to the subjectivity of the sense, the understanding does not have to rely on rational thinking, but can be built on an emphatic or even artistic-receptive base. Analogously, social action is for its part a special form of action. The specialization is manifested in the associated sense of a social action, which has to refer to the behaviour of other individuals. In this way, the meaning of a social action is comprised of the interactions between individuals in a social system. Figure 1.2 embodies a visualization of the specialization from behaviour to social action ([26, 27]: p. 4).

Actions, taken by agents of an enterprise, can be equally conceived as social actions, since they establish a relationship to other agents in the social system of the enterprise. In the context of enterprises, two perspectives contribute to the formation of the meaning of actions. On the one hand, social actions in an enterprise relate to co-workers and colleagues. On the other hand, an agent in an enterprise has to meet the obligations given by his superiors, which also influences his actions. In every enterprise, agents have to understand the actions of superiors first, before they can take their own social actions (managing their day-to-day business). The actions of superiors will specify schedules and requirements, to which operative actions have to conform. By fulfilling these requirements, the performing agents inevitably have

Fig. 1.2 Theory of social
action with reference to [26]

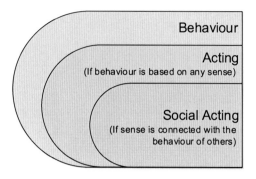

to rely on interaction with other agents, making it necessary to adapt their social actions to the behaviour of others.

Among other issues, this peculiarity is not considered by the theory of social action, as it solely refers to a subjective sense. Due to the general applicability of the theory, it misses a specific context, for which its explanative power leads to a possible explanation regarding modern enterprise arrangement and therefore misses a connection to modern enterprise culture. Specifically, the interaction between individuals and technology is mostly neglected [3], such as the interaction between individuals through technology [9]. The latter represents one of the most powerful manners for limiting the respective actions of an individual and therefore has to be considered in a modern theory of social action, as these considerations lead to new innovations for enterprises [6], but do have further an extensive impact on society.

However, the meaning of a social action within an enterprise is formed through consideration of given requirements and through the necessary social interactions between individuals. Acting in an enterprise therefore involves limiting the possible actions by respecting the given superior directives. The consideration of both superiors and those, who interact noticeable within a social system, follows from the theory of social actions. However, since this separation is not valid in every social system, it demands the introduction of an extended theory describing enterprise actions as a special form of social actions.

1.3   The Theory of Enterprise Action

1.3.1   Acting in an Enterprise Environment

The social system of an enterprise is a special one. It can be described as structural, hierarchical and functional system ([23]: p. 14). Every enterprise comprises different agents, which are related to each other. In addition these agents are in a hierarchical relation to each other. Classically those agents involved in planning tasks are superior to those conducting these tasks. From a functional perspective, the enterprise ultimately has to consider its varying stakeholders, such as the customers and competitors. With this representation of an enterprise, the actions of the agents have to be oriented towards the enterprise strategy. For the underlying value of an action therefore one has to consider both the social values of the agent, as well as the requirements of him as an agent in an enterprise. The fulfilment of these requirements is in the end motivated through the desire to rise the own economical capital, represented by the wage. The following Fig. 1.3 illustrates the conceptualization of enterprise actions.

The predefined actions for an agent are usually economically motivated by an enterprise. Nonetheless, taking such actions can also be based on cultural or social values, if these values are encouraging the respective participation in such enterprise or the commencement of the given predefined actions. However, when considering individuality at taking actions, conflicts with the own values can arise. Those conflicts have to be solved with creativity [15]. Two customer consultants for example have to increase the number of contract formations in their domain to fulfil the requirements of their enterprises. But since they have different personalities, one of them might try this in an unobtrusive manner, while the other one pursues a rather brisk strategy of convincing customers of the formation of a contract. Both of them are exposed to the values of the enterprise, i.e. to maximize the customer base and both of them will try to accomplish this objective differently according to their own values and the respective social context.

As a consequence we identify an individual problem of the agents to harmonize the values of the social system, i.e. the enterprise, and the own values. The enterprise culture then emerges from the manifold of single problems of harmonization of the different agents. In the way that the social actions are in reciprocal exchange with the enterprise culture, there is also a reciprocal connection between the enterprise culture and the enterprise actions. An enterprise action is thereby defined as any behaviour that has to be oriented towards one's own social values, as well as to the

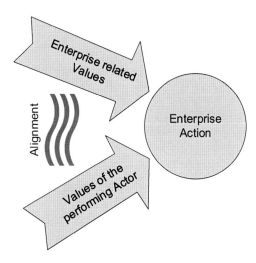

Fig. 1.3 Conceptualization of enterprise actions

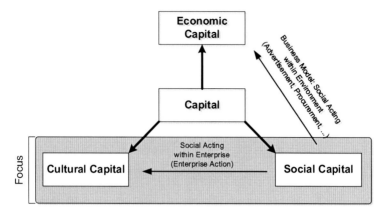

Fig. 1.4 Forms of capital with reference to [4]

predefined values of the enterprise. The purpose of enterprise actions is to increase economic, social and/or cultural capital (c.f. Fig. 1.4) [4] on the basis of predefined and personal values.

Although the predefined values of the enterprise are most likely oriented towards monetary magnitudes or economical capital, the emergence of a positive enterprise culture does not have to be solely accounted for by the personal values of single agents (or only by the predefined values of the enterprise). The predefined values can be either beneficial or harmful for the enterprise culture. Analogously, the values of the agents can be either beneficial or harmful, since their resulting actions influence other agents. Altogether both predefined and personal values should complement and balance each other with the aim of maximizing economic, social and cultural capital together harmonically. If the predefined and personal values only focus on the maximization of economical capital for example, the social and cultural capital would be neglected in the long term. As illustrated in Fig. 1.4, the presented research abstracts from any economic considerations. This is due two reasons, firstly, the examination of how economic capital can be gained through social capital is generally considered by business model research, e.g. [2, 16, 20]. Additionally, such research is considered with questions of market positioning [21] and therefore disregards mostly personal values, because while enterprise culture is external to a single individual, it is not external to the enterprise itself. Secondly, it is assumed that a structured figuration of culture, implicitly affects the economical chances of an enterprise, but initially requires a kernel theory, which is proposed by this paper.

1.3.2   On the Specification of Enterprise Actions

Actions will be specified as plans by the enterprise management [8], e.g. as Balanced Scorecard [17], to provide a basis for the agents to execute the actual actions. With

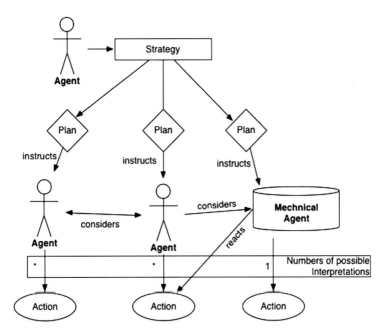

Fig. 1.5 Communication of the enterprise related values through the provision of plans for the execution of actions by agents

this specification of the actions for the agents, the management is able to orient the actions towards the predefined strategy of the enterprise. Thus such a plan is the instrument to communicate, on a more specific level, the business strategy and thereby the enterprise related values that serve as an orientation for the enterprise actions (cf. Fig. 1.5).

What is needed in the process of specification is a differentiation into plans for either human or automatic recipients. While the human agent, capable of interpreting a plan with respect to his personal experiences, is able to deal with potential room for interpretations (in possibly different manners) [14], the automated actions need strict requirements and guidelines for their execution. Thus for an action that is supposed to be executed by a mechanical agent, the management needs to specify a detailed description. This description is manifested in both a conceptual description from the management and at last in the software systems or more specifically their program code, which executes the action in the end [19]. Therefore the plan for human agents can and will leave room for interpretation, while the plan for a mechanical agent is precisely specified. This explicit relationship between an automatic action and the original plan can be seen in Fig. 1.5. The human agent however, given his own non-static interpretations, has always the possibility of adapting his actions. For even with increasingly specific plans, the enterprise will have no influence on the covert behaviour of the human agents.

1.3.3 The Search for a Cultural and Enterprise Value of Actions

From the perspective of the enterprise actions, the automatic actions are always only oriented towards the enterprise related values. This is the case, since the provided descriptions of the actions do not have to be aligned with the personal values of the agent. Actions executed by human agents on the other hand receive the personal values of the agents through room for interpretation. The human agent can decide how he wants to execute an action. The previously mentioned customer consultants still have the choice of how to advice the customers of their enterprise. Analogously, actions within an enterprise can be oriented towards the personal values of the human agents. Asking about the wellbeing of a person before inquiring about the processing state of an order can strengthen the social recognition of the questioned one. If this social recognition reversely reacts with other individuals, a respectful social intercourse can be maintained, which leads to an increase in the cultural capital of the enterprise. The social recognition of single agents thus becomes to a culture of mutual recognition within an enterprise.

A mechanical agent is not able to make situational decisions or diverge unexpectedly from its instructions. A specific diversion from the automated action is not possible, since automated enterprise actions rely solely on the enterprise related values, manifested in program code. With equal treatment of mechanical agents, qualified by the own task spectrum they have to work on, a reciprocal dependence between human and mechanical agents is created. Accordingly, human agents are dependent on machines and programs in the execution of their own day-to-day business. The communication has to be oriented towards the language of the mechanical agent at all times, since understanding is here statically defined. Through this orientation towards the mechanical agent, the human agent is denied to develop the own personality, gathering experience and in the end making personal progress, respectively building social capital and contributing to cultural capital. The language of the mechanical agent to receive commands and make orders has to become the language of the human actor, whose stasis can only be broken by external progression or by the exchange of the mechanical agent. A specific situation is visualized in Fig. 1.6, where a business process model is divided into three systems, each responsible for an individual activity of the model. Two of the three activities are already automated and only one activity still relies on the interaction with a human agent. The human agent receives his work instructions according to the process model from a mechanical agent, which provides the individual appointments of various customers in a well-defined format. These appointments then have to be prioritized. Although the human agent is able to perform this task freely according to his own creativity, as long as he conforms to the enterprise related values, in the end his decisions have to be well formalized for the following activity, where again an mechanical agent is at work.

Based on the well-defined formats for automated activities and the possibility for the management to instruct the mechanical agents precisely, there is no need for a

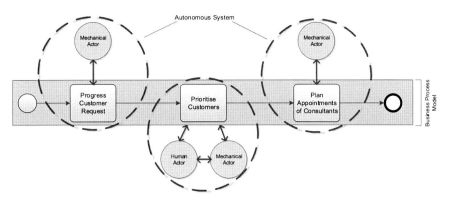

Fig. 1.6 Business process model with automated and partially automated activities

harmonization when only mechanical agents are present. Finding the best form of executing an action to combine enterprise related and personal values is no longer required, since there is now just one possible form of the action, which is completely oriented towards the enterprise related values. This issue can be subsumed by the upcoming Fig. 1.7, which includes two curves describing the possibility of actions and their reference to enterprise value by either the human or the mechanical actor. While it is possible, to let a mechanical actor learn over time, e.g. through machine learning [1], this benefit will stagnate over time, when no chances are left for improvement. Furthermore the possibility of machine learning is not necessarily available, for example by mechanical actors that are part of a production process, as these actors need to provide the identical predefined level of quality beginning with their activation. The second curve represents the possible development of actions by the human actor. With respect to efficiency, in those cases, where a mechanical actor is available, it will probably surpass the human. However, the human might be able to question a certain task and might adapt his actions accordingly. For example, the human could experience obsolete, but prescribed tasks and simplify them. If this adaptation comes with any value, it probably serves the enterprise. However, this innovative value decreases by the adaptation of competitors and the possibility of using a mechanical actor for performing these tasks. The latter could then again peak an innovation, but the human decision for using such a mechanical actor remains innovative, while the actions of the mechanical actor will again stagnate on a certain level of value. However, the value of an action by a mechanical actor is determined by the action of the human action of deciding for, developing of etc. these mechanical actors.

Consequently there is no cultural capital and no enterprise culture in an enterprise that has been subject to a complete automation. The formation of cultural capital is not possible in largely automated enterprises, since social capital cannot be converted to cultural capital with only few or even no possibilities of human interaction. Conclusively, the human action needs to be in continuous rebuttal in order to gain any value from any action within an enterprise. Therewith, enterprises

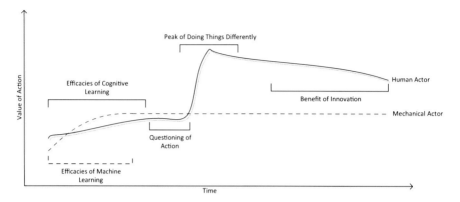

Fig. 1.7 Curves of both the correlation between the enterprise value and the action of the human as well as the machine

can only change with respect to their environment, if the human actions change. This defines the superorder of human actions to mechanical actions.

1.4 Conclusion

To illustrate the drawbacks of a progressing automation that focuses solely on economical capital, the continual loss and first of all the relevance of a culture could be explained. But instead the actually noticeable consequences of such automation for an enterprise should be investigated for the purpose of elucidating and conveying the relevance of this perspective. Enterprises need in the long term competitive advantages to position themselves in their respective context and to secure the existence of the enterprise [Por80]. Those competitive advantages differentiate the own enterprise from competitors on the market. A competitive advantage can only arise through human agents. Although there can be competitive advantages due to automated processes, this automation first has to be developed by a human agent. With the technological possibilities, e.g. due the unlimited number of software instances, competitive enterprises will adapt their processes to a similar or identical degree. Thereby, the initial advantage will only result in barriers for entering the market and the required differentiation at the market has to be settled based on other characteristics. Facing a complete automation, a consequence would be an exclusion from the market or the achievement of a quasi-monopoly [21], because without any cultural capital a further differentiation is not possible. For example, a manufacturer with a fully automated factory enters the market and acquires cost-, quality- and time-leadership. As soon as other enterprises copy or imitate this technology, the need arises to differentiate the enterprises on the basis of other criteria. Since mechanical agents are only capable of reacting to predictable conditions, human agents are needed to cope with the altering market. A differentiation and the viability of an enterprise are thus only given through human agents. Contrary to mechanical agents, the human agents of an enterprise cannot be identical to those of other enterprises and are therefore the only magnitude that is able to account for such differentiation. The lacking issue of harmonization for mechanical agents due to their absence of personal values will also miss a contribution to the security of the enterprise success. For that reason, in addition to the economic capital, the social capital and cultural capital are absolutely necessary for the competitive viability of an enterprise.

In the end it is a question of the responsibility that computer sciences have to bear, if they focus exclusively on the efficiency of an enterprise. We need methods or extensions of existing methods that facilitate a planning of enterprises beyond monetary magnitudes. Enterprise modelling for instance, in making fundamental assumptions about the social system of an enterprise [10], should consider the short-term nature of success that is purely measured in efficiency. Efficiency through progressing automation disregards the human being and can thus endanger the viability of the enterprise. Concluding systems design needs to respect any situation for which it encounters the exclusion of social interactions by means of replacing human agent by mechanical agents, whereby, it should not be threatened by specific humanistic consequences. Moreover systems design human interaction needs to be perceived as a key stone for planning and design information systems. Regardless if it is excluded or enabled the consequence have more facets than an economic perspective.

References

1. Alpaydin E (2004) Introduction to machine learning. MIT Press, Cambridge. ISBN 0262012111
2. Barone D, Jiang L, Amyot D et al (2011) Reasoning with key performance indicators. In: Johannesson P, Krogstie J, Opdahl AL (eds) The practice of enterprise modeling SE – 7. Springer, Berlin/Heidelberg, pp 82–96. doi:10.1007/978-3-642-24849-8_7. ISBN 978-3-642-24848-1
3. Boudreau M-C, Robey D (2005) Enacting integrated information technology: a human agency perspective. Organ Sci 16(1):3–18. doi:10.1287/orsc.1040.0103, INFORMS
4. Bourdieu P (2012) Ökonomisches Kapital, kulturelles Kapital, soziales Kapital. In: Handbuch Bildungs- und Erziehungssoziologie Bildung und Gesellschaft. o.V., pp 229–242
5. Curtis B, Kellner MI, Over J (1992) Process modeling. Commun ACM 35(9):75–90. doi:10.1145/130994.130998
6. Dakhli M, De Clercq D (2004) Human capital, social capital, and innovation: a multi-country study. Entrep Reg Dev 16(2):107–128. doi:10.1080/08985620410001677835, Routledge
7. Davenport TH, Short JE (1990) The new industrial engineering: information technology and business process redesign. Sloan Manag Rev 31(4):11–27
8. Ferreira A, Otley D (2009) The design and use of performance management systems: an extended framework for analysis. Manag Account Res 20(4):263–282. doi:10.1016/j.mar.2009.07.003
9. Fiol CM, O'Connor EJ (2005) Identification in face-to-face, hybrid, and pure virtual teams: untangling the contradictions. Organ Sci 16(1):19–32. doi:10.1287/orsc.1040.0101, INFORMS

10. Frank U (2012) Multi-perspective enterprise modeling: foundational concepts, prospects and future research challenges. Int J Softw Syst Model. doi:10.1007/s10270-012-0273-9

11. Gibb AA (1987) Enterprise culture – its meaning and implications for education and training. J Euro Ind Train 11(2):2–38, doi:10.1108/eb043365, MCB UP Ltd. ISBN 10.1108/eb043365

12. Hammer M, Champy J (2006) Reengineering the corporation: a manifesto for business revolution. Revised Up. HarperBusiness, New York. ISBN 0060559535

13. Heinrich L (2001) Wirtschaftsinformatik, 2nd edn. Oldenbourg, München/Wien, p 380

14. Joas H (1993) Pragmatism and social theory. University of Chicago Press, London. ISBN 0226400425

15. Joas H (1996) The creativity of action. University of Chicago Press, Chicago, p 352

16. Kaplan RS, Norton DP (2000) Having trouble with your strategy? Then map it. In: Focusing your organization on strategy – with the balanced scorecard, 2nd edn. Harvard Business School Publishing Corporation, Boston

17. Kaplan RS, Norton DP (1996) The balanced scorecard. Harvard Business School Press, Boston

18. Lee EA (2008) Cyber physical systems: design challenges. In: 2008 11th IEEE international symposium on object and component-oriented real-time distributed computing (ISORC). IEEE, pp 363–369, doi:10.1109/ISORC.2008.25. ISBN 978-0-7695-3132-8

19. Leonardi PM (2011) When flexible routines meet flexible technologies: affordance, constraint, and the imbrication of human and material agencies. MIS Q 35(1):147–167

20. Osterwalder A, Pigneur Y (2010) Business model generation: a handbook for visionaries, game changers, and challengers. Wiley, Hoboken. ISBN 0470876417

21. Porter ME (1980) Competitive strategy: techniques for analyzing industries and competitors. The Free Press, New York

22. Rajkumar R (Raj), Lee I, Sha L et al (2010) Cyber-physical systems: the next computing revolution. In: Proceedings of the 47th design automation conference on – DAC '10. ACM Press, New York, NY, USA, p 731, doi:10.1145/1837274.1837461. ISBN 9781450300025

23. Ropohl G (1978) Einführung in die allgemeine Systemtheorie. In: Lenk H, Ropohl G (eds) Systemtheorie als Wissenschaftsprogramm. Athenäum, Königstein

24. van der Aalst WMP (2004) Business process management: a personal view. Bus Process Manag J 10(2):135–139

25. van der Aalst WMP, Ter Hofstede AHM (2003) Business process management: a survey. In: Proceedings of the 1st international conference on business process management, volume 2678 of LNCS. Springer, Berlin/Heidelberg/New York, pp 1–12

26. Weber M (1978) Economy and society. University of California Press, Berkeley/Los Angeles/London. ISBN 0520035003

27. Weber M (1922) Gesammelte Aufsätze zur Wissenschaftslehre. J. C. B. Mohr, Tübingen

Chapter 2
The Human Factor in Computer Science and How to Teach Students to Care: An Experience Report

Janet Siegmund and Sven Apel

Abstract The human factor plays a crucial role in software engineering, so software engineers should pay sufficient attention to it. In this paper, we present our experience with teaching software-engineering students to care about the human factor. In particular, we report on a course that we conducted at the University of Magdeburg, in which we applied explorative and interactive techniques to teach the basics of human-behavior measurement. In summary, we received mostly positive feedback of the students, and found that after the course, students are able to properly take care of the human factor.

2.1 Importance of the Human Factor

In the late 1960s, software developers had to face increasingly complex software, eventually leading to the software crisis. In part, the crisis was caused by the fact that software was not developed for humans, but for computers. As Dijkstra phrased it in his 1972 Turing lecture "The humble programmer"[1]:

> [O]ne programmer places a one-line program on the desk of another and either he proudly tells what it does and adds the question "Can you code this in less symbols?" [...] or he just asks "Guess what it does!"

In these days, programming was seen as art—understandability or maintainability of source code was not the primary concern. Furthermore, usability of programs was not an issue, because only few, highly trained people worked with computers. Today, almost everyone uses computers regularly, for example, when using a smart phone. Even globally dispersed team members can collaborate on a single project

---

[1] http://dl.acm.org/citation.cfm?id=355604.361591

J. Siegmund (✉) • S. Apel
University of Passau, Passau, Germany
e-mail: siegmunj@fim.uni-passau.de; apel@uni-passau.de

© Springer International Publishing Switzerland 2014
K. Zweig et al. (eds.), *Socioinformatics - The Social Impact of Interactions between Humans and IT*, Springer Proceedings in Complexity,
DOI 10.1007/978-3-319-09378-9_2

with the help of collaborative software systems. Thus, the role of humans, either alone or as a group, either as a developer or as a user, is important.

Unfortunately, human behavior is difficult to predict; we cannot easily predict whether two humans in the same situation behave the same—we cannot even predict whether one human behaves the same if encountering the same situation twice. Instead, we need to conduct empirical studies, in which we observe people when they work with source code or when they use a program, so that we can predict how new programming languages or tools or features of a program affect the human who is using it.

However, conducting empirical studies in software engineering is rather uncommon. Only recently, empirical investigations, especially those that focus on the human factor, become more and more common. Before that, researchers who have developed new techniques with the goal of improving comprehensibility of source code or the usability of user interfaces, often argued only with plausibility arguments about why the technique or interface is more comprehensible or more usable. In practice, the claimed benefits may not hold or are difficult to evaluate.

For example, in Word 2000, Microsoft introduced adaptive menus.[2] Instead of a fixed order, menu items arrange according to their frequency of usage, so their order changes during usage. This way, the designers hoped to increase the efficiency of using Word, because frequently used menu items were always on top. However, in practice, this did not work, because with adaptive menus, the location of menu items appears to be non-deterministic. Thus, users look for a menu item in the wrong place, and are slower with adaptive menus.

One reason for the reluctance of conducting empirical investigations is that they require considerable effort and knowledge, which is often not taught during computer-science education. Thus, researchers often underestimate the effort and importance of a sound study design, or depend on trained experimenters.

To address this problem, we need to train software-engineering researchers in empirical methods. Although the call to integrate empirical methods in the computer-science curriculum is not new [2, 4], empirical methods are only in few universities part of the software-engineering curriculum, for example, at the Karlsruhe Institute of Technology (Walter Tichy), Freie Universität Berlin (Lutz Prechelt), or University of Marburg (Christian Kästner).

We designed and held a course at the University of Magdeburg, which was initially based on course developed by Christian Kästner at the University of Marburg, in cooperation with Stefan Hanenberg. For our teaching philosophy, we changed several aspects of the course, including the order and emphasis of the material for our course design. In this paper, we report our experience with teaching empirical methods to computer scientists, for which we adopted explorative and interactive teaching methods. Our contribution in this paper is twofold:

---

[2]http://www.nofluffjuststuff.com/blog/aza_raskin/2007/03/are_adaptive_interfaces_the_answer_/

- We share our experience with teaching experience a course on empirical methods, including comments of enrolled students.
- We discuss the contents and teaching methods of our course to help other researchers and teachers in designing a course similar to ours.

Our overarching goal is to raise the awareness of the human factor in software-engineering education as well as software engineering in general. Furthermore, we want to initiate the path toward a common course description and teaching material, so that software engineers can properly address the human factor in software engineering.

2.2   Content of the Course

In this section, we present the content of the teaching course, which we held at the University of Magdeburg in the winter term 2012/2013, and which we will hold at the University of Passau in the summer term 2014. In the course, we have covered software measurement, the importance of the human factor, systematic planning and conducting of experiments, as well as quantitative and qualitative methods.

2.2.1   Software Measurement

In this part, we have covered typical computer-science topics, that is, topics, that the typical student has encountered during his/her education at the University of Magdeburg. We have started with measuring software based on software measures, because that is close to students' previous experience. We have introduced different software measures, such as lines of code and cyclomatic complexity, and let students compare different software systems based on software measures. During the measurement process of software, typically no unsystematic measurement error occurs, that is, the same software, measured with the same tool, always has the same software measures. In this context, the measurement process is straightforward and intuitive.

As next step, we introduced performance measurement, that is, students should measure the performance of software systems. This introduces the concept of measurement error, as measuring the same software system under the same condition does not necessarily lead to the same results. Thus, to measure one software system, several measurement runs should be conducted (which is what all students intuitively did). Furthermore, we did not specify performance further, leaving the operationalization to the students.

2.2.2   The Human Factor

Based on these first steps, students have experienced that even without the human factor, having sound measurement procedures already requires considerable effort. When the human factor is added, there is even more variation. For example, different developers have different levels of programming experience, are familiar with different programming languages or domains, and might prefer one tool over the other. Thus, different individuals introduce measurement bias, which has to be taken care of.

Hence, with this first part, we have shown students the necessity for sound methodological training in empirical research. Even for empirical investigations without the human factor, such as performance measurements, significant bias can be introduced. In our course, we continued with introducing a structured way to conduct empirical research, which we present in the next section.

2.2.3   Conducting Empirical Investigations Systematically

To minimize bias in empirical investigations as much as possible, we need a structured way to conduct them. To this end, we can divide empirical investigation into five stages: objective definition, design, conduct, analysis, and interpretation. Next, we give a short overview of each stage.

Objective Definition.   First, during the objective definition, the goal of the empirical study is defined. This includes stating research hypotheses or questions and *operationalizing* the constructs of interest. The research hypotheses drive the further design and prevent "fishing for results", that is, playing around with the data until something interesting is found.

Experimental Design.   Second, we need to develop the experimental design, which defines how we evaluate the hypotheses or how we answer the questions. A major obstacle in this stage is to control for *confounding parameters*, which may severely bias the results. In our experience, handling confounding parameters is the most difficult and most often neglected part of empirical studies, because researchers simply are not aware of them.

Conduct.   Third, the experiment is conducted. In this phase, despite all careful planning, numerous things can go wrong. For example, experimenters can influence participants, participants deviate from their instructions, or there might be power failures. All deviations should be thoroughly documented.

Analysis.   Fourth, the data obtained from the experiment conduct needs to be analyzed. In our experience, researchers often do not know what to do with data beyond computing average scores or visualizing data. However, average scores are not sufficient to answer the question whether a difference is real or whether data correlate. To this end, we need to conduct statistical tests. Furthermore, we teach standard visualization techniques, such as box plots and histograms, which is important to present the results.

Interpretation.   Last, we need to interpret the data, which goes beyond accepting or rejecting hypotheses or answering research questions. Instead, we need to state what the results mean. Is a new programming paradigm better for programming experts, but not for novices? Should we really start teaching programming with this programming paradigm? Thus, the results need to be put into perspective beyond the experiment.

In our course, we taught students to follow this procedure when designing experiments. Next, we introduced quantitative and qualitative methods to measure the human factor.

2.2.4   Quantitative and Qualitative Methods

To measure the human factor, there are different quantitative and qualitative methods. For a better overview, we developed a mindmap for students, and showed them at the beginning of each lecture, where the topic belongs to. In Fig. 2.1, we show the map.

We started with quantitative methods, because in discussions with fellow researchers and students, we learned they are more intuitive for computer-science students than qualitative methods. Specifically, we discussed how to design

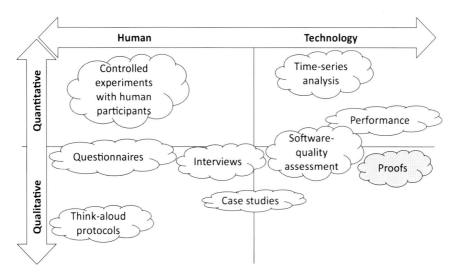

Fig. 2.1 Mindmap on qualitative and quantitative methods in conjunction with the human/technology factor for the empirical-methods course. Elements with a *gray background* were omitted in the course, but shown for completeness

controlled experiments, in which several participants are observed, and in which only few data points per participants are measured. For example, we explained how to measure program comprehension based on response time and correctness for a group of 20 or more participants. In this part, we also introduced typical analysis procedures and their logic, from descriptive statistics (e.g., mean, standard deviation, box plots, confidence intervals), to significance testing (e.g., χ^2 test, Student's t test, ANOVA).

After quantitative methods, we introduced qualitative methods, in which few participants are observed in detail. In particular, we introduced interviews, case studies, and think-aloud protocols. Furthermore, we discussed analysis techniques for qualitative data, from grounded theory to card-sorting techniques, including reliability measures, such as Cohen's Kappa for interrater reliability.

With teaching these techniques, we enabled students to select an appropriate empirical method for an evaluation of the human factor, so that students are able to conduct an empirical study in their Bachelor's or Master's thesis.

For the course design, we relied on our experience when discussing empirical studies with colleagues and students. In particular, we often found that people need some time to grasp the difficulties of soundly measuring the human factor, so we decided on the order that we felt is most intuitive for computer-science students. Furthermore, we selected methods that we often encountered in the literature, so that students learned a representative set of empirical techniques.

While in this section, we concentrated only on the content we taught, we explain in the next section the teaching methods, that is, *how* we ensured that students understand and apply suitable empirical methods.

2.3 Teaching Methods

Teachers face the problem of getting students to deeply understand and care about the contents of a course. In Fig. 2.2, we show the six levels of learning, as described by Bloom and others [1]. At the lowest level (*knowledge/remembering*), students take in data, remember it, and recite it. At the second level (*understanding/comprehension*), students can explain information in their own words. At the third level (*application*), students are able to apply information in a new way. Students at the fourth level of learning (*analysis*) can break information down into its parts, and at the fifth level (*synthesis/creation*), can combine the parts to build a new structure. At the highest level (*evaluation*), students can judge the value of information. In our course, we aim at the highest level of learning, *evaluation*, so that students can select an appropriate method for any empirical question.

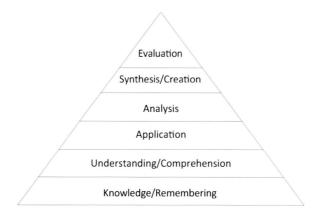

Fig. 2.2 Levels of learning according to Bloom [1]

So, how do we achieve the highest level of learning? In our course, we used a combination of explorative and interactive teaching methods, which we explain in the next sections.[3]

2.3.1   Exploration

During the complete course, we let students explore empirical methods for themselves, with the goal that they notice the problems when using an intuitive approach. For example, regarding performance measurement, we divided the students in groups of three to four students each and let them evaluate the run-time performance of different sorting algorithms in different programming languages. We did not explain to students how performance is reliably measured, or how to control for confounding parameters, but we let them follow their intuition. Afterwards, the student groups compared their results, and found that no group would trust the results of another group. We concluded this exercise with a guided discussion that led to the conclusion that we need sound empirical methods to control for confounding parameters, so that we get trustworthy and practically applicable results.

Another explorative method was that students read and discussed papers that report controlled experiments. Based on these examples, students learned how to address typical problems of empirical research (e.g., how to operationalize variables,

---

[3]There are several books (e.g., [5]) and online sources (e.g., University of Zurich (http://www.hochschuldidaktik.uzh.ch/hochschuldidaktikaz.html)) that give comprehensive overviews of explorative and interactive teaching methods.

how to control for confounding parameters), as well as how typical pitfalls can be avoided.

To ensure the highest level of learning, students conducted their own empirical study. To this end, students could select any question they were interested in and design an experimental plan to evaluate the question. After feedback on the experimental plan (to ensure that students would not run into too much trouble), students conducted the experiment, analyzed the data, and wrote a report (available on the course's website: http://wwwiti.cs.uni-magdeburg.de/iti_db/lehre/emcs/2012/). Thus, students experienced themselves the complete process of one empirical study, which we believe showed them the difficulty of sound empirical studies, and also sensitized them for the importance of the human factor.

2.3.2   Interaction

Another feature of our course is interactive teaching, that is, involving students in the lecture, not just presenting them information. In addition to exploration, a lot of interactive methods exist. That includes asking students during the lecture to suggest solutions to presented problems, were appropriate, after a few-minutes discussion with their neighbor (referred to as *buzz groups*).

To highlight how to involve students in the teaching process, we discuss how we introduced the systematic procedure to conduct empirical investigations (cf. Sect. 2.2.3). For brevity, we focus only on the first stage. First, we introduced a running example, that is, we used the research question *Do comments make source code more comprehensible?* as starting point. Then, we started with the objective definition, in which we introduced the terms independent and dependent variable, as well as hypothesis. After explaining what an independent variable is, we asked students to name the independent variable in the example (which is *comments*). We explained what operationalization is, and then asked students to operationalize comments (e.g., comments can either be present or not present, or comments can be incorrect or correct). We did the same for the dependent variable (i.e., *comprehension*), which can be operationalized with letting participants fix a bug, and then measure the correctness or response time of the bug fix. The faster the response time, the better comprehension should have taken place. After defining the variables, we talked about hypotheses, and that they must be falsifiable, that is, if they are wrong, we must be able to show that. We asked students to decide whether the research question is suitable as hypothesis (which it is not, because it is too unspecific), and then, after a short discussion with their neighbor, give examples of more suitable hypotheses (e.g., *Incorrect source-code comments slow down program comprehension*). For the remaining stages, we used the same pattern, that is, explaining information to students for 5–10 min, and then asking them to apply the information right away. This way, they deepened their understanding and can better memorize information, which is a well-studied phenomenon [3].

We also included other interactive methods, of which we present a couple of examples, with which we had a particularly good experience in our course. First, we used *black stories/situation puzzles* as interactive methods. Situation puzzles are a game, in which a host explains a situation to the players, and the players have to find out by asking only yes-or-no questions how this situation emerged. In the course, we presented students the conclusion of an experiment, for example, that expert and novice programmers show equivalent program comprehension, and students should come up with the experiment plan. In this case, there were different source-code snippets, and several of them contradicted the expectation of expert programmers, making them as incorrect and slow as novice programmers [6]. This way, students learned to look at experiments from a different angle, which we believe helped them to get a deeper understanding.

As second example, we used the interactive methods *world café*, *vernissage*, and *student award*. In the world café, students designed an empirical study in a group of about four students for a given research question (e.g., *How do students learn programming?*), and prepared a flip chart to present their results to the other groups, which was then presented (vernissage). This way, students applied the taught techniques and deepened their understanding in discussions. After the vernissage, students selected the best experimental design (student award), which helped them to critically review and understand the experimental plan of the others.

To summarize, with explorative and interactive teaching methods, we ensured a higher level of learning, compared to the classical way of merely presenting the information to students. And this is also what our students said in the evaluation, which is conducted for every lecture (see next section).

2.3.2.1 Evaluation

To ensure high-quality teaching, it is custom at the University of Magdeburg to conduct an evaluation at the end of the semester. In this evaluation, students can give comments about what they like and dislike about a course, and how it can be improved. We received ten answers for our course, which can be summarized as follows:

1. First, we found that most of the students liked the interactive teaching style, because it motivated them to analyze the presented information actively (instead of only listening to information).
2. Second, most students also liked that they could explore information based on real examples, and that they could try out the empirical methods and analysis techniques on short examples as well as on their own project.
3. Third, in contrast to the positive feedback, there are also few students who did not like the explorative part, especially when they had to conduct their own study.

Despite the negative feedback, we feel confident that our students profit from this way of teaching. Thus, when we conduct the course again, we will continue using the explorative and interactive teaching techniques.

Considering the projects of the students, we observed a surprisingly high quality of studies. For example, one project explored factors influencing personal web-search behavior. To this end, the students assessed several factors that might influence web-search behavior, such as experience with using the web, frequency of using computers or the web, or education, and designed web-search tasks of different difficulty levels. In the analysis, students correlated the performance in the web-search tasks with the assessed factors, and created hypotheses based on their analysis. During all experimental stages, students used the proper methods and draw the correct conclusions from their results. The complete report is available on the project's website (http://wwwiti.cs.uni-magdeburg.de/iti_db/lehre/emcs/2012/projekte/FabianAnton.pdf).

Considering the positive evaluation results and the high quality of the empirical studies, we believe that our course prepares computer-science students well to soundly measure the human factor in computer science.

2.3.2.2 Course Description

Most university courses require a course description as part of the examination regulations for a study course. In Table 2.1, we show how we described the course as example for other teachers.

We suggest to offer the course to graduate students, because the course requires reading and understanding scientific papers, which is often too advanced for undergraduate students. Furthermore, we can plan the course for 5 or 6 ETCS, which translate into 60 h of attending the lecture and participating in experiments, and 90 (5 ETCS) or 120 (6 ETCS) hours of homework assignment, including the project. Since in the project, students apply the taught techniques on their own study, the time students invest for the project should not be reduced, but only the time for the homework assignment. We recommend that students are familiar with software engineering, because most of the examples are in this domain. This course description is intended to help other researchers integrate a similar course at their university.

Table 2.1 Excerpt of the syllabus for the empirical-methods course

Issue	Content
Course name	Empirical methods for computer scientists
Level	Master
Teaching	4 h per week
Effort	5/6 ETCS
Recommended prerequisites	Knowledge on software engineering
Learning goals	After this course, students: • Know empirical methods for evaluating research questions • Can assess the validity of scientific statements • Can apply a suitable empirical method for evaluating research questions in a bachelor's or master's thesis
Content	Results in computer science often aim at better quality, lower costs, better maintainability, or better comprehensibility. To be able to evaluate these claims, we need to use empirical methods, which are the content of this course. For illustration, we use examples from software engineering and programming languages. Contents include: • Scientific method, proofs, empirism • Rigorous measurement of performance, including benchmarks, case studies • Controlled experiments with developers • Statistical background knowledge
Relevant for examination	• Participation in lectures • Participation in experiments of other students of this course • Evaluating a self-selected research question • Completing homework assignments • Oral examination

2.4   Summary

In this paper, we highlighted the importance of the human factor in software engineering. We argued that one reason for the underrepresentation lies in the negligence of empirical methods in the computer-science curriculum. We shared our experience from our course we held at the University of Magdeburg, so that other researchers are encouraged to introduce a similar course at their university. This way, we hope to have raised the awareness for the necessity of teaching empirical methods to computer scientists, and to have helped other researchers to introduce a similar course at their university. The material of the course is available online (http://wwwiti.cs.uni-magdeburg.de/iti_db/lehre/emcs/2012/).

Acknowledgements Thanks to Christian Kästner and Stefan Hanenberg for fruitful discussion on the design of the course. This work has been supported by the German Research Foundation (AP 206/4, AP 206/5, and AP 206/6).

References

1. Bloom B, Engelhart M, Furst E, Hill W, Krathwohl D (1956) Taxonomy of educational objectives: the classification of educational goals. Handbook 1: cognitive domain. David McKay, New York
2. Braught G (2005) Teaching empirical skills and concepts in computer science using random walks. In: Proceedings of technical symposium on computer science education (SIGCSE), St. Louis. ACM, pp 41–45
3. Craik F, Tulving E (1975) Depth of processing and the retention of words in episodic memory. J Exp Psychol 104(3):268–294
4. Joint IEEE Computer Society/ACM Task Force for CC2001 (2001) Computing Curricula 2001. http://www.acm.org/education/curric_vols/cc2001.pdf
5. Silberman M (1996) Active learning: 101 strategies to teach any subject. Pearson
6. Soloway E, Ehrlich K (1984) Empirical studies of programming knowledge. IEEE Trans Softw Eng 10(5):595–609

Chapter 3
Socially-Aware Traffic Management

Michael Seufert, George Darzanos, Ioanna Papafili, Roman Łapacz, Valentin Burger, and Tobias Hoßfeld

Abstract Socially-aware traffic management utilizes social information to optimize traffic management in the Internet in terms of traffic load, energy consumption, or end user satisfaction. Several use cases can benefit from socially-aware traffic management and the performance of overlay applications can be enhanced. We present existing use cases and their socially-aware approaches and solutions, but also raise discussions on additional benefits from the integration of social information into traffic management as well as practical aspects in this domain.

3.1 Introduction

In online social networks (OSNs) users voluntarily provide information about themselves, their interests, their friends and their activities, especially about their current situation or exceptional events. Additionally, other usage data might be recorded clandestinely. Nowadays these so called social signals are ubiquitous and can not only be collected from OSNs (e.g., friendships, interests, trust-relevant metadata), but also from applications (e.g., messaging or call patterns) and sensors (e.g., location). Social awareness harvests these signals, extracts useful and re-usable information (e.g., users' social relationships, activity patterns, and interests), and exploits them in order to improve a service.

M. Seufert (✉) • V. Burger • T. Hoßfeld
Institute of Computer Science, University of Würzburg, Würzburg, Germany
e-mail: seufert@informatik.uni-wuerzburg.de; valentin.burger@informatik.uni-wuerzburg.de;
hossfeld@informatik.uni-wuerzburg.de

G. Darzanos • I. Papafili
Department of Informatics, Athens University of Economics and Business,
Athens, Greece
e-mail: ntarzanos@aueb.gr; iopapafi@aueb.gr

R. Łapacz
Poznań Supercomputing and Networking Center, Institute of Bioorganic Chemistry
of the Polish Academy of Sciences, Poznań, Poland
e-mail: romradz@man.poznan.pl

© Springer International Publishing Switzerland 2014
K. Zweig et al. (eds.), *Socioinformatics - The Social Impact of Interactions between
Humans and IT*, Springer Proceedings in Complexity,
DOI 10.1007/978-3-319-09378-9_3

Recently in the field of traffic management in the Internet, works were conducted which utilize social information, for example, to avoid congestion, increase bandwidth, or reduce latency. In that context, social awareness links social signals and information, such as social network structure, users' preferences and behaviors, etc. to network management mechanisms. This means that such mechanisms exploit the information in order to perform efficient network management, content placement, and traffic optimization to enhance the performance of an overlay application (e.g., video streaming, file sharing). As this promising research field has yet got little attention, this work will provide an insight to this new topic.

In Sect. 3.2, we will define socially-aware traffic management and present the involved actors. In the remainder of the paper, we focus on use cases in which the involved stakeholders can benefit from social information. In Sect. 3.3, we present three use cases (i) content storage and delivery, (ii) service mobility, and (iii) network security, in which social awareness can benefit involved stakeholders. Finally, Sect. 3.4 raises discussions on additional benefits from the integration of social information into traffic management and concludes this work.

3.2 Terminology, Definitions, and Actors

In order to put the description and discussion of socially-aware traffic management and its use cases on a firm footing, we start with definitions of important terms and present the involved actors.

3.2.1 Terminology and Definitions

Social signals are any signals which are emitted by persons. In the special case of Internet services, we consider a social signal to be a signal which is emitted in the Internet by an end user of an Internet application. Thus in fact, any interaction of an end user with an Internet service is a social signal. A signal itself contains no information, but information can be created out of them when evaluated in the right context. As signals range from simple logins to a service to complex service requests which might include interactions with other users or the environment, examples are manifold. In the context of online social networks, these signals are, e.g., friendship requests and confirmations, indications of interest or liking, or postings about external events. Another example are location data which are created by sensors of mobile devices and are transmitted when using an Internet service.

Social information is defined as information about one or more persons, or their relationships. It is deduced from bringing social signals into an appropriate context which allows for the generation of new insights about respective users or relationships between users. For example, evaluating the signals that user A sent a friendship request to user B, and B's confirmation of that request, will generate the social information the A and B are friends. Evaluating the same signals differently (i.e., in a different context), will give the information that A and B were online and used the OSN service at the time the signals were emitted. As another example, evaluating the location data signals of user C in the context that every second Saturday the location data is the same, and that there is a football stadium at that specific location, will create the information that C is a supporter of a certain football team. Thus, it can be seen that the created information depends on the particular evaluation of the social signals, and might require additional (external) information in order to create new social information. Usually, such partial information which requires external information to generate new social information is called meta information.

In general, the term **social awareness** implies the utilization of social information for a specific purpose. In the context of Internet services, we will consider social awareness to be the utilization of social information to improve an Internet service. Social awareness can include the collection of social signals and production of social information, but also a collaboration with a social information provider (see below) is possible. Taking provided or generated social information as an input, social awareness will exploit this information in order to deliver a higher service quality to end users and/or to provide the service more efficiently.

Socially-aware traffic management is a special case of social awareness, in which social information is used to improve traffic management on the Internet. Traffic management are means in order to handle the transportation of data across the networks. As not only link capacities increase in the Internet, but also traffic volumes become larger due to new service levels and new applications (e.g., cloud applications), it is necessary in order to avoid congestion, deliver applications in acceptable quality, and to save energy, resources, and costs. Traffic management can be employed by the service itself, e.g., by service quality selection or scheduling of transmissions, or by the network operators. Their methods typically include, but are not limited to, prioritization, routing, bandwidth shaping, caching, or offloading. The utilization of social information shall enable the improvement of classical traffic management solutions as well as the development of novel traffic management approaches.

3.2.2 Actors of Socially-Aware Traffic Management, Their Goals, and Possible Benefits

With socially-aware traffic management, five actors and their goals have to be considered. Note that each actor can be a separate stakeholder, but stakeholders can also have multiple roles.

The **cloud service provider** or **application provider** provides an Internet service to end users. The offered service might be running on own infrastructure or on the infrastructure of a cloud operator. The application provider is interested in monetization of offered services which includes reduction of Internet service provider (ISP) infrastructure and cloud resource consumption costs. Satisfaction of end users is a crucial issue as it is directly related to the number of customers. To ensure this goal the Quality of Service (QoS)/Quality of Experience (QoE) requirements should be met [8]. If social information is utilized, QoS/QoE parameters for services may be improved and also new services may be offered. Moreover, infrastructure costs can be reduced if social information is exploited to increase the utilization of resources.

The **datacenter operator** or **cloud operator** is operating a datacenter/cloud infrastructure. He offers cloud resources, e.g., storage, computation, to the application provider, while he buys Internet connectivity and inter-connectivity of his sites from an ISP. The cloud provider is mainly interested in monetizing his infrastructure and reduce his costs. Monetization of infrastructure is done by fulfilling service level agreements (SLAs) with the application provider and therefore guaranteeing satisfactory QoS parameters for end users. Reduction of costs, in case of cloud providers, focuses on best possible utilization of hardware – both resource-wise and energy-wise. As it depends on ISPs to provide network access, this stakeholder will seek the best SLA conditions for himself.

The **Internet service provider (ISP)** is operating a communication network infrastructure. His main interest is monetizing his infrastructure. This can be increased by high quality of network services that translates into satisfaction of cloud operators and also of end users [12]. Supporting new services, possibly by employing social information, may be attractive for application providers, and simultaneously makes the ISP more competitive towards end users and cloud providers. Such new services can also enable reduction of costs by both more efficient use of own resources and keeping transit link traffic as low as possible.

The **end user**'s main concern is his own QoE [8], network access cost, and energy consumption [13]. This stakeholder is rather not involved in other stakeholders' interactions, being primarily a client of ISP and application provider. It is noteworthy, that costs in case of end user often can be expressed by being exposed to advertisements instead of being involved in the monetary flow.

The **social information provider** wants to benefit from his social information. Therefore, he can provide or sell social information to application providers or ISPs in order for the latters to support optimization decisions, e.g., content placement.

3.3 Use Cases for Socially-Aware Traffic Management

The exploitation of social information may lead in significant benefits for all involved stakeholders, i.e., ISP, cloud operator, application provider and the end user. Therefore, three indicative use cases are presented. For *content storage and delivery*, three variations of socially-aware traffic management are described, that are centralized, distributed, or hierarchical content delivery platforms. Moving towards practical applications, we investigate information spreading in OSNs and its employment in socially-aware caching solutions for video streaming, and we overview existing traffic management solutions that employ social signals to perform efficiently content placement or pre-fetching. Next, we describe the *service mobility* use cases, which involves WiFi offloading, content placement for mobile users, and service placement. Finally, we provide some insight to a third use case, i.e., *network security* employing social information to defend against Sybil and DDoS attacks.

3.3.1 Content Storage and Delivery

Internet traffic has increased manifold in the last few years. Drivers for this increase include inter alia the increased popularity of video streaming applications (e.g., YouTube, NetFlix), the emergence of a multitude of new overlay applications such as online storage (e.g., Dropbox, Google Drive) and online social networks (e.g., Facebook, Twitter), the high increase of mobile devices (e.g., smartphones, tablets) and the upcoming trend of moving both storage and computing capacity to the cloud which allows more, even smaller players to enter the market [4]. Concerning video as a key application contributing largely to the overall IP traffic, video and specifically user-generated content (UGC) sharing (e.g., home-made videos) has evolved to a major trend in OSNs. Three variations of the content storage and delivery use case are described where social information is employed to achieve efficiency in content delivery, in terms of either content placement or pre-fetching. Moreover, we present work in literature which analyze content spreading in OSNs and show already existing socially-aware caching solutions for video streaming. Finally, we briefly overview related works which employ social awareness in order to handle the huge traffic volumes generated by video sharing over OSNs.

3.3.1.1 Exploitation of Social Information by a Centralized Content Delivery Platform

We consider a use case inspired by the evaluation scenario described in Traverso et al. [24]. Specifically, we consider an OSN having users around the globe who share videos via the OSN which are stored in third-party owned online video streaming platform such as YouTube. This content can be viewed by their online friends, their friends' friends, etc. through the Friend-of-Friend (FoF) relationship.

In order to meet the content demand by users of the video streaming platform, who are located worldwide, the video platform is operated on a geo-diverse system comprising multiple points-of-presence (PoPs) distributed globally. These PoPs are connected to each other by links, which can either be owned by the entity that also owns the PoPs, or be leased from network providers. Each user is assigned and served out of his (geographically) nearest PoP, for all of his requests as depicted in Fig. 3.1.

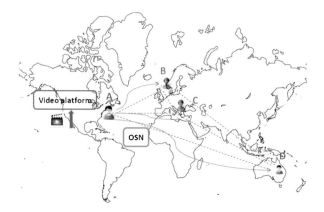

Fig. 3.1 Content delivery in geographically distributed PoPs (Source: inspired by Traverso et al. [24])

Placing data close to the users is an approach followed by most content delivery networks (CDN). Therefore all content uploaded by a user A is first uploaded to the nearest PoP, i.e., PoP_A. When content is requested by another user B, the nearest PoP to B, i.e., PoP_B, is contacted and if the content is available there, the request is served. The content can be already present at PoP_B, if content was first uploaded there or was brought there by an earlier request. If the content is not available in PoP_B, then a request is made to PoP_A and the content is brought to PoP_B.

In such as setup, social relationships between the users of the OSN can be taken into account to predict where a piece of content will be requested next, i.e., by which PoP. For instance, it is expected that due to their social relationship, users of the online social network will request a piece of content that a friend of them, e.g., user A, has uploaded to the video platform with higher probability than users that have no social relationship with A.

The so-called social awareness involves the exploitation of such social information (e.g., social graph, social relationships among users, behavior patterns, login time, time spent in the OSN) in order to predict where and by whom an uploaded video will be consumed. Such predictions can be employed to develop socially-aware mechanisms such as TailGate proposed in Traverso et al. [24] that will enable pre-fetching of the uploaded video in various locations (e.g., PoPs).

3.3.1.2 Exploitation of Social Information by a Distributed Content Delivery Platform

An alternative use case involves the dissemination of video content in a peer-to-peer (P2P) fashion among an OSN's users, taken from Li et al. [15]. We consider again an OSN, whose users are scattered around the globe and upload videos, i.e., UGC to an online video streaming platform like YouTube. This content, similarly to the scenario described in the previous section, can be viewed by their friends, their friends' friends, etc.

End users, also called peers, download parts of the file, e.g., chunks or blocks, and are considered to be able to re-upload them to another peer. Additionally, a proxy server is considered to orchestrate the content dissemination as a P2P tracker or to participate in it. In the latter case, the proxy server is connected to the content provider, which is an end user in case of UGC. Moreover, multiple proxy servers are considered to be also distributed globally and each one of them to have a specific domain of influence, e.g., an ISP's domain, an Autonomous System (AS).

The initial content provider uploading a video to the proxy server, the proxy server itself, and the peers participating in the dissemination of that particular video are considered a swarm. Furthermore, the video parts exchange among peers is performed based on some specific peer and chunk selection policy. As mentioned before, placing video chunks close to the end users is an approach followed by most CDNs as it leads to lower latency and stall time, and thus high QoE for end users. Therefore, social information can be extracted from OSN by the video platform owner, so as to predict by whom a video uploaded to the proxy server will be viewed. These users can be preferably included in the dissemination swarm. Once they want to access the video, they have lower delay (thus, a better QoE) because part of the file is already on their device.

According to Traverso et al. [24] and Li et al. [15], direct friends (1-hop friends) and friends of friends (FoF or 2-hops friends) of a user C have high probability (more than 80 %), to watch a video uploaded or posted by C. Social information, such as users' interests, e.g., sports or music, prove to be also important, as users, which have a FoF relationship with C and share the same interests, are highly likely to watch a video uploaded by C.

3.3.1.3 Exploitation of Social Information by a Hierarchical Storage Platform

Another interesting use case of applying the knowledge derived from OSNs is improving the internal decision making algorithms in advanced distributed hierarchical storage management systems.

Hierarchical Storage Management (HSM) is an approach to manage high volume of data in the way that data are categorized and moved between storage types to reduce the storage cost as well as to optimize an access and the energy consumption of data storage management. A hierarchy level is assigned to a storage media. The first level is represented by high-speed high-cost devices destined for data set that is frequently accessed by applications. Other data, for example older and thus less popular, can be automatically moved to a slower low-cost storage media.

Usually, three levels of storage hierarchy are defined, as illustrated in Fig. 3.2. The first level is a high-speed systems, such as hard disk drive array, the second one is slower, such as optical storage, and the last one may be implemented as magnetic

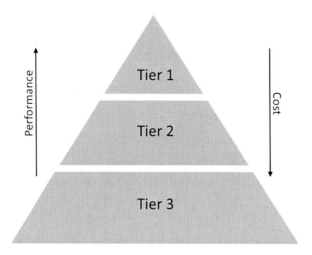

Fig. 3.2 Storage tiers in HSM along with performance and cost trends (Source: inspired by Ganley [9])

tape drives. As the technology of the first level is the most expensive the size of it is smaller than the storage sizes of other levels.

HSM can be also interpreted as a tier storage technique, although sometimes storage specialists see differences between them [29]. The basic difference seems to be the way how datasets are accessed. In case of HSM, inactive data are moved to the levels of slower storages and can be accessed directly again only after migrated back to the first high-speed level. On the contrary, the tier approach allows fetching data from any tier any time.

Nowadays, when the amount of data is rapidly growing, HSM offers a substantial benefit from managing storage devices efficiently, especially in large-scale networks, storage and computational environments, such as clouds. In particular, a common deployment scenario involves resources of a cloud residing in remote geographical locations, while end users perceive its resources as a consistent pool available for allocation (e.g., IaaS model [17]). This operation is highly related to the so-called service mobility discussed in Sect. 3.3.2.

Moreover, one cloud operator may utilize storage resources or assign specific works to another cloud operator, e.g., in the context of a cloud federation, to achieve e.g., load balancing, reduction of his individual energy consumption, etc. In order to optimize these operations of data migration, social signals can be exploited by the cloud operators so as to predict not only the amount but also the where and when of future demand. As a result, end users will experience better QoE, i.e., faster access to data, while the cloud operators will achieve more accurate utilization of their storage hierarchies (tiers) and in consequence lower energy consumption.

3.3.1.4 Understanding Information Spreading in OSN
for Utilization in Traffic Management Algorithms

Social awareness can be used in different ways to improve content delivery. To exploit the information within OSNs it is important to understand how information is spread, and how to identify important nodes in the social graph and the relationships with their friends.

In Bakshy et al. [2], the authors investigate the influence of posts by tracking the diffusion of URLs in Twitter and show that content that is connected with good feeling and interesting content is more likely to be propagated. They also find that the users that have most influence are also the most cost-effective. Hence, influential users post relative rarely, but if they do, the content is of high interest.

In Ruhela et al. [19], the authors collected data from five different sources and investigated the temporal growth and decay of topics in the network and the geographical and social spread of the topics. Besides identifying different classes of temporal growth patterns and time zone differences in popularity, they find that the social cohesion of users interested in specific content is greater for niche topics. Hence, they propose to use semantic information about the topic to assess the temporal growth, use time-zone information to predict the breakout of popular topics in a specific region, and use social network predictors for niche content. To distribute the content and use cache capacities effectively we need good replica placement algorithms.

Next, in Wittie et al. [28], the authors inferred the network structure of Facebook performing crawling, packet captures, and network measurements. Due to high locality of interests they state that service providers could profit a lot from locality to save traffic on intercontinental paths. Proposed solutions are regional caches or a CDN that connects a global network of server farms at different ISPs to bring the content close to users.

Finally, in Wang et al. [27], the authors explore how patterns of video link propagation in a microblogging system are correlated with video popularity on the video sharing site, at different times and in different geographic regions. Then, they design neural network-based learning frameworks to predict the number and geographic distribution of viewers, in order to deploy a proactive video sharing system. The evaluations show that their frameworks achieve better prediction accuracy compared to a classical approach that relies on historical numbers of views.

3.3.1.5 Existing Socially-Aware Caching Solutions for Video Streaming

Socially-aware caching tries to predict future access to user generated content (e.g., videos) based on information from OSNs. Hints shall be generated for replica placement and/or cache replacement.

In Sastry et al. [21] the classical approach of placing replicas based on access history is improved. Therefore social cascades are identified in an OSN, and declared affiliations of potential future users (i.e., OSN friends of previous users) are

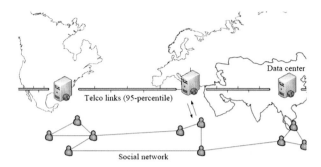

Fig. 3.3 TailGate's generic distributed architecture for video delivery (Source: Traverso et al. [24])

added. In Scellato et al. [22] standard cache replacement strategies are augmented with geo-social information from OSNs. Again social cascades are analyzed to recognize locally popular content which should be kept longer in the cache.

Apart from the increasing popularity of video sharing over OSN, another significant characteristic of content dissemination on top of OSNs that need to be taken into consideration is the long-tailed nature of content, i.e., UGC such as home-made funny videos, etc. Below, some solutions are briefly presented and discussed that focus and address the long-tailed nature of video delivery over OSNs.

In Traverso et al. [24], the authors propose TailGate which derives and uses social information derived from OSNs, such as social relationships, regularities in read access patterns, and time-zone differences for predicting where and when the content will likely be consumed, in order to push the content where-ever before it is needed. Thus, exploiting the derived social information, long-tail content is selectively distributed across globally spread PoPs, while lowering bandwidth costs and improving QoE. In particular, bandwidth costs are minimized under peak based pricing schemes (95th percentile), but the approach is also beneficial for flat rate schemes.

For the analysis of TailGate, the authors considered the scenario depicted in Fig. 3.3. Specifically, they consider an online video delivery service with users across the world, operated on a geo-diverse system comprising multiple PoPs distributed globally. Each of these interconnected PoPs handles content for geographically close users. In particular, when UGC is created, it is first uploaded to the geographically closest PoP, and then it can be distributed to other PoPs.

In Li et al. [15], the authors identify the fact that the deployment of traditional video sharing systems in OSNs is costly and not scalable. Thus, they propose SocialTube, a peer-assisted video sharing system that explores social relationships, interest similarity, and physical location between peers in OSNs. Specifically, SocialTube incorporates three algorithms: an OSN-based P2P overlay construction algorithm that clusters peers based on their social relationships and interests, an OSN-based chunk pre-fetch algorithm to increase the video prefetch accuracy to minimize video playback startup delay, and a buffer management algorithm. The social network-based P2P overlay has a hierarchical structure that connects a

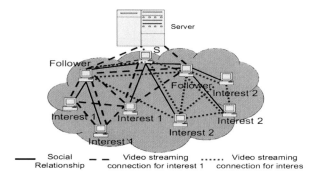

Fig. 3.4 SocialTube's P2P overlay structure based on social relationships and interests (Source: Li et al. [15])

source node with its followers, and connects the followers with other non-followers (Fig. 3.4).

Moreover, in order to reduce the video startup latency, the social network-based pre-fetching algorithm is employed. This algorithm dictates that when a source node uploads a new video to a centralized video server, the source also pushes the prefix, i.e., the first chunk, of the video to its followers. Additionally, it is pushed to the peers in the interest clusters matching the content of the video, because there is a high probability that it will be requested to be watched, since followers watch almost all videos of the source.

In Zhou et al. [31], the authors examined crawled data from Facebook and observed that a significant fraction of Internet traffic contains content that is created at the edge of the network, i.e., UGC. Moreover, they observed that users are in general significantly more interested in the content that is uploaded by their friends and friends-of-friends, while traffic local to a region is produced and consumed mostly in the same region, which is contrary to the case of traditional web content. Furthermore, they argue that while caching the most popular 10 % of traditional content would allow to satisfy at least half of all requests, this caching technique would perform significantly worst for content with a more even popularity distribution.

Therefore, they propose WebCloud, a content distribution system for OSNs that works by re-purposing client web browsers to help serve content to others, and which tries to serve the request from one of that user's friends' browsers, instead of from the OSN directly. WebCloud is designed to be deployed by a web site, such as the provider of an OSN, to be compatible with the web browsers (no plug-ins) of today and to serve as a cache for popular content.

The authors claim that WebCloud trying to keep the content exchange between two users within the same ISP and geographic region, to reduce both OSN's and the ISP's costs. WebCloud emulates direct browser-to-browser communication by introducing middleboxes, which are called redirector proxies (Fig. 3.5). The proxy determines if any other online local user has the requested content and if so, fetches

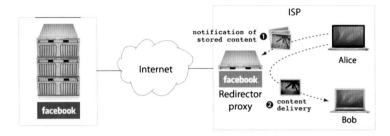

Fig. 3.5 Content sharing in WebCloud, where Alice first informs the proxy of locally stored content. When Bob requests content from the proxy, the proxy fetches it from Alice and delivers it to Bob, thereby keeping the content exchange local (Source: Zhou et al. [31])

the content from that user's browser and transmits it to the requestor. Should no local user have the content, the browser fetches the content from the OSN.

3.3.2 Global Service Mobility

In today's Internet, services and applications have to be seamlessly available to end users at any time and location. Especially mobile users, i.e., users who access services by means of mobile devices from any location, pose severe challenges for mobile service providers, cloud operators and application providers.

3.3.2.1 Exploitation of Social Information for WiFi Offloading and Service Placement

New services have ever increasing network requirements which strain current mobile networks, and feed the desire of network operators to offload traffic to WiFi networks. On the other hand, application providers rely more and more on the cloud concept, which allows moving services within or among datacenters worldwide, and thereby foster the mobility of services. Both approaches can benefit from the utilization of social information, as well as data related to the location of the end user.

Location information can be retrieved easily either from the mobile network provider, from services (e.g., [19]), or from end users. Location data are often used by applications, and even shared by end users, e.g., as meta-information of postings, or as explicit postings of locations in OSNs or in specialized services like Foursquare.[1]

---

[1] https://foursquare.com

This allows for the monitoring of such social signals and facilitates the creation of mobility patterns for different users. One step beyond, these patterns can be used to predict the location of a user in the future. These predictions can be also used to improve the content delivery described above in Sect. 3.3.1 for mobile users. Both pre-fetching and caching algorithms will achieve a higher accuracy which improves the cache hit ratio for mobile users, and thus, improves users' QoE. In the following we will present other aspects of global service mobility, such as WiFi offloading and service placement.

Offloading data to WiFi networks has already been in the focus for some years, and can be considered as providing fast and energy-efficient Internet access for mobile end users. Offloading traffic using WiFi networks can save between 75 and 90 % of the energy for network transmissions compared to 3G connectivity only [10]. At the same time, the risk of network caused stallings may be reduced by increased data rates on WiFi, improving the QoE of the end user.

Service placement is a generalization of content placement, i.e., instead of only placing content at the appropriate caching locations (cf. Sect. 3.3.1), whole services are placed. This includes the creation, termination, and migration of virtual machines which are running the service. Especially cloud services which are based on the elasticity of clouds can benefit from such placement. The description is mainly based on Biancani and Cruschelli [3].

Service placement is interesting both from ISPs' and application providers' perspectives. Application providers are interested in maintaining a good ratio of revenue and costs. An optimal placement of a service among a number of cloud providers or own datacenters can help optimizing costs as well as meeting end users' QoE requirements. However, the optimal placement of services is also in the interest of the ISP to reduce his operational costs, as a disadvantageous service placement can increase traffic from outside the provider's AS. Thus, both stakeholders could collaborate, e.g., by using an ALTO [1] style approach.

The placement of service will be optimized by taking into account social information, e.g., where services may be popular in specific regions or for which specific groups. Such information can be aggregated from different sources, i.e., from a direct cooperation with OSN applications and end users, or from the ISP who might exploit its aggregated knowledge on users interests and mobility patterns. A side effect of these new possibilities is an expected improvement of the perceived network quality on end user devices. This is achieved by locating services closer to the end user, reducing the delay, and improving the network throughput.

3.3.2.2 Existing Socially-Aware Solutions for Service Mobility

Fon[2] started a WiFi sharing community in 2006 by offering a home router device with a separate shared WiFi network which could be accessed by every community

---

[2]http://www.fon.com

member. Similar approaches are the hotspot databases Boingo[3] and WeFi,[4] and Karma[5] which adds social reciprocity to WiFi sharing. Also the research community investigated incentives and algorithms for broadband access sharing [16], and architectures for ubiquitous WiFi access in city areas [20, 26].

Valancius et al. [25] propose a distributed service platform, called Nano Data Centers or NaDa, based on tiny (i.e., nano) managed "servers" located at the edges of the network, i.e., in users' premises. With NaDas, both the nano servers and access bandwidth to those servers are controlled and managed by a single entity, typically an ISP. The role of the nano servers can be played by ISP-owned devices like Triple-Play gateways and DSL/cable modems that sit behind standard broadband accesses. Such gateways form the core of the NaDa platform and can host many of the Internet services today operated in datacenters. ISPs can easily employ NaDas by providing new customers with slightly over-dimensioned gateways, whose extra computation, storage, and bandwidth resources are used to host services, all of which will be totally isolated from the end user via virtualization technologies.

Home router sharing based on trust (HORST) [23] is a mechanism which addresses the data offloading use case and combines it with mechanisms for content caching/pre-fetching and content delivery. HORST establishes a user-owned Nano Data Center (uNaDa) on the home router and sets up two WiFi networks (SSIDs) – one for private usage and one for sharing. The owner of the home router shares the WiFi credentials with trusted users via an OSN application and can also request access to other shared WiFis. As HORST knows the location of the users and the WiFis, it can recommend near shared WiFi networks, and automatically request access and connect the users for data offloading. HORST combines content placement for mobile users with data offloading and uses social information in order to predict which content will be requested by which user. As HORST also knows about the current and predicts future users of each shared WiFi from location data, the uNaDa on the home router can be used to cache or pre-fetch delay-tolerant content which will be delivered when the user is connected to the WiFi.

QoE and Energy Aware Mobile Traffic Management (QoEnA) [14] is a mechanism focused on the improvement of QoE, at the same time reducing the energy consumption on mobile devices by intelligent scheduling of network traffic which is generated on the mobile device. It is based on QoS maps, user mobility prediction, energy models, and QoE models. Thereby, QoEnA schedules traffic which is generated on the mobile device to different connections or locations in order to improve the QoE of the end users while reducing the energy consumption of the mobile device.

Social information can also be used for routing and content placement in mobile ad hoc networks. In Costa et al. [5], a routing framework for publish-subscribe

---

[3]http://www.boingo.com

[4]http://wefi.com

[5]https://yourkarma.com

services is described. In such a service, messages (or content items) are tagged with topics and shall be routed to users that are interested in these topics. In the presented framework, predictions of co-locations are based on metrics of social interactions, because socially bound hosts are likely to be co-located regularly. For each message, a best carrier is selected based on interests, mobility, and co-location prediction, to whom the message is forwarded. The presented socially-aware approach is shown to have advantages in terms of message delivery, delay, and overhead.

Dinh et al. [6] work towards socially-aware routing for mobile ad hoc networks. They present an algorithm to identify modular structures in dynamic network topologies based on interactions, and merge them to a compact representation of the network. This compact representation is well suited for dynamic networks and allows for a faster computation of routing strategies compared to state-of-the-art algorithms.

3.3.3 Network Security

OSNs can provide valuable social information that can be employed to come up against malicious users and their behavior which leads to large scale attacks, e.g., sybil or DDoS attacks which are described below.

Social information about trust between users can be used both for the self-protection of the OSN and for the protection of other services or applications. This information can be either extracted from the OSN itself, or by creating a graph of trust among the end users of a service or application, i.e., by employing a system in which each user has the ability to create relationships of trust with other users.

In both approaches, users can be represented as nodes of a social graph where an edge between two nodes implies a both-way relationship of trust.

A sybil attack [7] occurs when a malicious user takes on a large number of identities and pretends to be multiple, distinct users/nodes. When these sybil nodes collude together and comprise a large fraction of systems identities, the attacker gains significant advantage in a distributed system. For example, sybil nodes can work together to distort reputation values, out-vote legitimate nodes in consensus systems, or corrupt data in distributed storage systems.

In order to avoid sybil attacks, the social activity of various nodes as well as their social relationships can be examined in order to verify fake profiles and identify potential malicious nodes in a system. Based on the fact that a social network is fast mixing [18] the social graph can help to reveal malicious users, while this becomes easier as the number of fake (malicious) identities increases. This is due to the fact that it is difficult for a malicious user to establish multiple social relationships between the sybil nodes and real users.

According to Yu et al. [30], sybil nodes form a well-connected subgraph that has only a small number of edges connected to honest users, as depicted in Fig. 3.6. These edges are also called attack edges to the honest network.

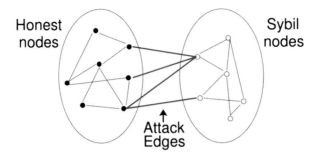

Fig. 3.6 The complete social graph consisting of the honest network, the sybil network, and the attack edges connecting these two networks (Source: Yu et al. [30])

As a counter-measure, SybilGuard [30] exploits this property of the social graph to identify sybil nodes by finding this small cut and by bounding the number of sybil nodes a malicious user can create. SybilGuard relies on a special kind of verifiable random walk in the graph and intersections between such walks. These walks are designed so that the small cut between the sybil region and the honest region can be recognized and used to identify malicious users.

A Distributed Denial of Service attack (DDoS) is another use case, which occurs when multiple systems usually controlled by one malicious entity (e.g., botnets) flood the bandwidth or resources of a targeted system, e.g., a server, in order to make the system unavailable to its intended users.

A similar solution to SybilGuard can be developed against DDoS attacks. The social graph can be used by OSNs or third-parties (e.g., video streaming platforms, or banking institutions) to reveal fake profiles and identify potentially malicious users. In a DDoS attack, a malicious entity may be in control of multiple systems belonging to real users, e.g., by means of trojan horses, and therefore making it unable to detect these problematic profiles through the social graph.

We can overcome this obstacle by observing/monitoring the social behavior/ activity of end users; whenever significant changes appear in their activity (e.g., high increase of requests for video viewing), the user should be added in a "suspects' list" implying that some or all of his requests are being denied. Additionally, the same solution can be used in applications or systems which do not contain social information. This can be achieved by asking users to sign up with a social network account or by encouraging them to declare other users in the system that they trust or are socially connected to. Maintaining such a suspects' list, we may avoid or relieve the impact of a DDoS attack.

Nonetheless, there are open issues to be addressed by future research. For example, there is the possibility that also honest users are included in the suspects' list, and thus get denied of the service. Moreover, there is a trade-off between the efficiency of the monitoring of the social behavior of users against the high monitoring effort.

3.4 Conclusions and Discussions

The research field of socially-aware traffic management opens new perspectives for improved service delivery in the Internet like the discussed use cases content storage, global service mobility, or network security. Nevertheless, the utilization of social information introduces new interdisciplinary research challenges, such as the monitoring of social data, the processing and storage of these data, as well as the integration into existing systems. From the network traffic management perspective, it is unclear how to realize and deploy socially-aware traffic management solutions. In particular, the design and implementation of socially-aware networking functionalities involves several stakeholders of the service delivery chain, like the social information provider, Internet service providers, cloud operators, application providers, and finally the end user. Hence, there must be an incentive-compatible network management mechanism [12] which satisfies the requirements of involved stakeholders (e.g., high QoE for end users, low traffic/congestion in ISPs' links, lower energy consumption in datacenters where services run), and which is based on well-defined open protocols as currently defined by the IETF ALTO "Application-Layer Traffic Optimization" Working Group in Alimi et al. [1]. Furthermore, a seamless integration of those socially-aware mechanisms into today's Internet applications and network management is desired. Such architectural and conceptual challenges are currently developed in the FP7 SmartenIT[6] for a tighter integration of network management and service functionality to offer a large business potential for all players involved. An initial architecture with incentives as integral part is presented in Hausheer and Rückert [11] which follows a modular concept and existing standards and proposals.

From a social data analysis perspective, there are also some practical aspects addressed in order to obtain, maintain, and update social signals from existing platforms due to the huge amount of existing data. While social data can be utilized for various use cases, the monitoring of the data differs in terms of temporal and spatial scale. Different sources for social information can be taken into account and it has to be decided which information to be retrieved from which source and when. Depending on the actual use case, it may be necessary to consider aggregated information, single or selected users, or even the entire OSN topology. Moreover, the monitoring frequency may address different timescales (hours, days, months). In general, there is a trade-off between accuracy and costs of social information, which may be adjusted by appropriate (temporal and spatial) sampling methods. This means that the social monitoring has to be customized for a specific data source from which relevant data is fetched. Then, preprocessing, aggregation, and analysis of the data is necessary before sending the resulting social information to traffic management elements in a system. Especially the design of such algorithms,

---

[6]The FP7 project SmartenIT (FP7-2012-ICT-317846) "Socially-aware Management of New Overlay Application Traffic combined with Energy Efficiency in the Internet" is running from Nov 2012 to Oct 2015. More information is available at http://www.smartenit.eu

e.g., to identify relevant nodes in the network responsible for video cascades, and the computational complexity have to be addressed, since scalability is one of the key issues of socially-aware traffic management. Finally, besides those technical challenges, privacy is another major challenge which has to be ensured and integrated in the solution space.

For overcoming the emerging challenges, the tight coupling between social data analysis and the resulting traffic management solutions is required, while socio-informatics is foreseen as a driver to establish an interdisciplinary research community in that interesting domain.

Acknowledgements This work was partly funded by Deutsche Forschungsgemeinschaft (DFG) under grants HO 4770/1-1 and TR257/31-1, and in the framework of the EU ICT Project SmartenIT (FP7-2012-ICT-317846). The authors alone are responsible for the content.

References

1. Alimi R, Penno R, Yang Y (2013) ALTO protocol. Technical report, Internet engineering task force application-layer traffic optimization working group. http://tools.ietf.org/wg/alto/
2. Bakshy E, Hofman JM, Mason WA, Watts DJ (2011) Everyone's an influencer: quantifying influence on Twitter. In: Proceedings of the 4th ACM international conference on web search and data mining (WSDM '11), New York
3. Biancani M, Cruschelli P (eds) (2013) Deliverable D1.2 report on cloud service classifications and scenarios, SmartenIT Consortium (European FP7 STREP No. 317846)
4. Cisco (2012) Cisco visual networking index: forecast and methodology, 2011–2016. Technical report, Cisco
5. Costa P, Mascolo C, Musolesi M, Picco GP (2008) Socially-aware routing for publish-subscribe in delay-tolerant mobile ad hoc networks. IEEE J Sel Areas Commun 26(5):748–760
6. Dinh TN, Xuan Y, Thai MT (2009) Towards social-aware routing in dynamic communication networks. In: Proceedings of the 28th IEEE international performance computing and communications conference (IPCCC), Phoenix
7. Douceur JR (2002) The Sybil attack. In: Peer-to-peer systems. Springer, Berlin/Heidelberg pp 251–260
8. Fiedler M, Hossfeld T, Tran-Gia P (2010) A generic quantitative relationship between quality of experience and quality of service. IEEE Netw 24(2):36–41
9. Ganley B (2013) Optimize the virtual desktop experience through strong back-end design. Technical report, Dell Power Solutions. http://i.dell.com/sites/doccontent/business/solutions/power/en/Documents/ps4q13-20130371-ganley.pdf
10. Gautam N, Petander H, Noel J (2013) A comparison of the cost and energy efficiency of prefetching and streaming of mobile video. In: Proceedings of the 5th workshop on mobile video (MoVid '13), New York
11. Hausheer D, Rückert J (eds) (2013) Deliverable D3.1 report on initial system architecture, SmartenIT Consortium (European FP7 STREP No. 317846)
12. Hoßfeld T, Hausheer D, Hecht F, Lehrieder F, Oechsner S, Papafili I, Racz P, Soursos S, Staehle D, Stamoulis GD, Tran-Gia P, Stiller B (2009) An economic traffic management approach to enable the TripleWin for users, ISPs, and overlay providers. In: Towards the future internet – a European research perspective, Future internet assembly. IOS Press, US, pp 24–34. http://ebooks.iospress.nl/book/towards-the-future-internet
13. Ickin S, Wac K, Fiedler M, Janowski L, Hong JH, Dey AK (2012) Factors influencing quality of experience of commonly used mobile applications. IEEE Commun Mag 50(4):48–56

14. Kaup F, Hausheer D (2013) Optimizing energy consumption and QoE on mobile devices. In: Proceedings of the IEEE international conference on network protocols (ICNP 2013), Göttingen
15. Li Z, Shen H, Wang H, Liu G, Li J (2012) SocialTube: P2P-assisted video sharing in online social networks. In: Proceedings of the IEEE INFOCOM, Orlando
16. Mamatas L, Psaras I, Pavlou G (2010) Incentives and algorithms for broadband access sharing. In: Proceedings of the ACM SIGCOMM workshop on home networks, New Delhi
17. Mell P, Grance T (2011) The NIST definition of cloud computing. Technical report, Recommendations of the National Institute of Standards and Technology
18. Nagaraja S (2007) Anonymity in the wild: mixes on unstructured networks. In: Privacy enhancing technologies. Springer, Berlin/Heidelberg, pp 254–271
19. Ruhela A, Tripathy RM, Triukose S, Ardon S, Bagchi A, Seth A (2011) Towards the use of online social networks for efficient internet content distribution. In: Proceedings of the IEEE 5th international conference on advanced networks and telecommunication systems (ANTS), Bangalore
20. Sastry N, Crowcroft J, Sollins K (2007) Architecting citywide ubiquitous Wi-Fi access. In: Proceedings of the 6th workshop on hot topics in networks (HotNets), Atlanta
21. Sastry N, Yoneki E, Crowcroft J (2009) Buzztraq: predicting geographical access patterns of social cascades using social networks. In: Proceedings of the 2nd ACM EuroSys workshop on social network systems (SocialNets), Nuremberg
22. Scellato S, Mascolo C, Musolesi M, Crowcroft J (2011) Track globally, deliver locally: improving content delivery networks by tracking geographic social cascades. In: Proceedings of the 20th international conference on world wide web (WWW2011), Hyderabad
23. Seufert M, Burger V, Hoßfeld T (2013) HORST – home router sharing based on trust. In: Proceedings of the workshop on social-aware economic traffic management for overlay and cloud applications (SETM 2013), Zurich
24. Traverso S, Huguenin K, Triestan I, Erramilli V, Laoutaris N, Papagiannaki K (2012) TailGate: handling long-tail content with a little help from friends. In: Proceedings of the 21st international conference on world wide web (WWW2012), Lyon
25. Valancius V, Laoutaris N, Massoulié L, Diot C, Rodriguez P (2009) Greening the internet with nano data centers. In: Proceedings of the 5th international conference on emerging networking experiments and technologies (Co-NEXT '09), Rome
26. Vidales P, Manecke A, Solarski M (2009) Metropolitan public WiFi access based on broadband sharing. In: Proceedings of the Mexican international conference on computer science (ENC 2009), Mexico City
27. Wang Z, Sun L, Wu C, Yang S (2012) Guiding internet-scale video service deployment using microblog-based prediction. In: Proceedings of the IEEE INFOCOM, Orlando
28. Wittie MP, Pejovic V, Deek L, Almeroth KC, Zhao BY (2010) Exploiting locality of interest in online social networks. In: Proceedings of the 6th international conference on emerging networking experiments and technologies (Co-NEXT '10), Philadelphia
29. Yoshida H (2011) The differences between tiering and HSM. http://blogs.hds.com/hu/2011/05/the-differences-between-tiering-and-hsm-hierarchical-storage-management.html
30. Yu H, Kaminsky M, Gibbons PB, Flaxman A (2006) SybilGuard: defending against sybil attacks via social networks. ACM SIGCOMM Comput Commun Rev 36(4):267–278
31. Zhou F, Zhang L, Franco E, Mislove A, Revis R, Sundaram R (2012) WebCloud: recruiting social network users to assist in content distribution. In: Proceedings of the 11th IEEE international symposium on network computing and applications (NCA), Cambridge

Chapter 4
The Social Dimension of Information Ranking: A Discussion of Research Challenges and Approaches

Ingo Scholtes, René Pfitzner, and Frank Schweitzer

Abstract The ability to quickly extract relevant knowledge from large-scale information repositories like the World Wide Web, scholarly publication databases or Online Social Networks has become crucial to our information society. Apart from the technical issues involved in the storage, processing and retrieval of huge amounts of data, the design of automated mechanisms *which rank and filter information* based on their relevance (i) in a given context, and (ii) to a particular user has become a major challenge. In this chapter we argue that, due to the fact that information systems are increasingly interwoven with the social systems into which they are embedded, the ranking and filtering of information is effectively a socio-technical problem. Drawing from recent developments in the context of *social information systems*, we highlight a number of research challenges which we argue should become an integral part of a social informatics research agenda. We further review promising research approaches that can give rise to a *systems design* of information systems that addresses both its technical and social dimension in an integrated way.

4.1 Introduction

From a computer science perspective, the study of *information systems* has long been focused on *technical challenges* arising in the storage, management and processing of large amounts of information, as well as in the efficient retrieval of those pieces of information that match certain criteria. One possible reason is that most of the information systems which have traditionally been studied, like e.g. electronic commerce applications or expert systems, are *centrally managed*. In such centrally managed systems, strict policies regarding the quality, reliability and structure of stored information can be imposed. By this, the problem of identifying

I. Scholtes (✉) • R. Pfitzner • F. Schweitzer
Chair of Systems Design, ETH Zurich, Weinbergstrasse 56/58, 8092 Zurich, Switzerland
e-mail: ischoltes@ethz.ch; rpfitzner@ethz.ch; fschweitzer@ethz.ch

© Springer International Publishing Switzerland 2014
K. Zweig et al. (eds.), *Socioinformatics - The Social Impact of Interactions between Humans and IT*, Springer Proceedings in Complexity,
DOI 10.1007/978-3-319-09378-9_4

relevant information could effectively be deferred to users, by requiring them to provide reasonable query criteria. The emergence of global-scale information repositories like, e.g., the World Wide Web (WWW), scholarly publication databases or Online Social Networks (OSNs), has considerably changed this situation. Not only have these systems given rise to a superabundance of information; the fact that they are not centrally managed also comes with the challenge that much of the information is of questionable credibility and quality. As a result, in addition to mere technical issues of storing, processing and querying such large-scale systems, mechanisms which allow to *rank and filter information* based on their relevance or the reputation of their source have entered the focus of research. During the last decade, such ranking and filtering methods have become vital for the success of large-scale information repositories: One could hardly imagine a World Wide Web without *network-based ranking mechanisms* like those at work at the heart of search engines [5]. Similarly, the overabundance of information in other areas, like e.g. scholarly publications has led to the development of ranking and filtering methods, and personalised recommendation schemes that aim at assisting users in the identification of relevant information.

An important feature of most of today's knowledge spaces is that they are created, organized and consumed in a distributed, and collaborative fashion by groups of humans interacting on increasingly short time scales, a process commonly subsumed under the umbrella of *social computing* or *social information processing*. The questions how humans discover and navigate information spaces, why they create links between pieces of information, and by which collective processes certain pieces of information become popular or are seen as most relevant are crucial for models underlying ranking and filtering mechanisms. Not only do they affect the ability of individuals or organizations to retrieve information. They are also of prime importance for society as a whole since notions of relevance in networks of linked information (a) are increasingly influenced by social processes and (b) can be an important driver of social dynamics themselves. The resulting feedback between the social and the information layer of collaborative knowledge spaces questions to what extent current information ranking measures – although being computed algorithmically – can actually be seen as *objective*. As a consequence, the design of information systems and the definition of information ranking methods has a social, political and ethical dimension that is often underestimated even though it crucially affects our knowledge society [17, 20].

Although it is clear that the social and the information layer of collaborative knowledge spaces are inherently coupled and thus inseparable, the question how the ranking and retrieval of information is influenced by the structure and dynamics of the social systems that create and consume them has been addressed partially at most. In this chapter, we thus highlight some of the resulting research challenges and introduce research directions that seem suitable to address them.

4.2   The Emergence of Social Information Systems

According to the Encyclopaedia Britannica, in the context of computer science an information system is "an integrated set of components for collecting, storing, and processing data and for delivering information, knowledge, and digital products" [41]. This notion is focused on the technical components, i.e. the hardware, software, storage and communication technologies involved. From this perspective, the Internet, as well as the numerous Web-based services built on top of it, are globally distributed information systems that consist of millions of servers, that are connected via telecommunication facilities and exchange information via well-defined protocols. An important development that can be observed since roughly one decade and which has mainly been due to the widespread adoption of Web-based technologies, is that information systems are becoming increasingly *participatory* or *collaborative*: in modern systems users play an active role not only as information consumers, but also as producers, editors or reviewers. A particularly prominent and successful example is the online encyclopedia WIKIPEDIA, which is collaboratively written, edited and maintained by millions of contributing users. When applying the above definition of information systems to WIKIPEDIA, one observes that – due to its focus on technical components – it falls short of considering its most important component: the huge number of collaborating users and the mechanisms and processes by which they coordinate. In other words, it is not (only) the technical components that define modern information system like WIKIPEDIA, it is rather a unique combination of an underlying technical infrastructure – in the case of WIKIPEDIA the *MediaWiki* software – with the users and collective social processes that produce and evaluate information and link it to each other. By means of these processes, humans become an integral part of information systems, rather than being mere consumers. Similar examples for information systems in which humans and their social interactions are not only an important, but the crucial component, can be found. Online Social Networks (OSNs) like TWITTER, FACEBOOK and DIGG, or the Blogosphere are examples. But also the system of scientific peer review and publication is crucially shaped by the interactions of its users. Building on the engineering-centered definition given above, we thus extend the notion of an information system to include this social layer and introduce the following, more *systemic*, notion of a **social information system**:

> *A social information system is an ecosystem of hardware, communication technology, software services, and interacting humans that collects, stores, processes, evaluates and delivers information and knowledge.*

In a social information system, the interactions between its users are a crucial factor for its functioning. At the same time, in most systems these social interactions are mediated via online communication mechanisms provided by the underlying technical infrastructure or communication system. As such, the social interactionsin these information systems are – much like technical components, protocols and algorithms – at least to a certain extent influenced by engineers and designers.

The main purpose of any (social) information system is to satisfy the *information need* of its users. In the case of WIKIPEDIA, this can be the retrieval of an article containing a specific fact or reference the user was looking for. In Web search engines, it generally is a particular web page containing the information the user was searching for. In other systems, like Online Social Networks, the *need* of a user is generally not actively expressed in terms of a query. Here, information propagation rather follows a *push* model, i.e. information is proactively distributed. A general problem arising due to the overabundance of information, is that the volume of information that could potentially match the need or interests of a user is too large, thus requiring a selection what to display to the user. The utility of a social information system for users can be expressed as the extent to which this selection matches their need. In the following, we briefly introduce *information ranking, filtering and recommendation* mechanisms which are frequently applied to maximise the utility for users. In particular, we comment on the social dimension influencing these mechanisms, thus setting the stage for the challenges that will be introduced in Sect. 4.3.

4.2.1 Information Ranking: The Social Dimension

Most of today's large-scale information systems are *uncurated*, meaning that information (a) can be added by a vast number of users and (b) that this information is typically not required to follow a given semantic model, like, e.g., the use of a certain terminology, a given set of keywords or other semantic annotations. The retrieval of relevant information based on a search of keywords in such repositories introduces a number of challenges. First of all, differences in the use of terminology as well as language ambiguities limit the precision of search results. Secondly, due to the low barrier for users to enter information, it often is of uncertain quality and reliability. And finally, due to the vast amount of information, for almost any keyword the number of information pieces matching a search is too large to be explored by the user. These characteristics require to impose a *ranking* on the search results, which is ideally based on the likelihood that a given piece of information matches the particular information need of a user. The simplest possible approach one can think of, is to rank results according to the number of occurrences of the search term in a given document. Clearly, such a ranking is not optimal, as (a) the ranking of a document with respect to a given search term can easily be manipulated and (b) it totally neglects other dimensions of relevance, like, e.g., the trustworthiness or reputation of its source.

Decades before the World Wide Web became popular, various *hypertext environments* have been proposed which allow to easily cross-link information and which thus provide means to tackle these challenges. In particular, it has been argued that the link structures of such systems allow to infer knowledge about the semantics of information and to make statements about the reputation of a source based on the number of documents that refer to it [8]. These early works have laid the foundation not only for the popularity of the World Wide Web [3], but also for the widespread adoption of graph-based methods in the ranking of search results [11, 19] which eventually gave rise to popular Web search services like ALTAVISTA or GOOGLE [5]. Even though its importance for the ranking of search results in GOOGLE is nowadays widely overestimated, the PAGERANK algorithm is a particularly well-known example for such a graph-based algorithm. Here the reputation of a particular document is recursively computed based on the reputation of the documents that link to it [27]. Clearly, this is only one particular metric for the *centrality* or *reputation* of nodes in a graph and numerous other methods have been investigated during the last decades [26].

A particularly interesting aspect of these graph-based methods is that they inevitably make assumptions about the *semantics* of a link between two pieces of information. In the case of PAGERANK, where the reputation of a source is – to a certain degree – passed on to the documents to which it refers, the assumption is that the formation of a link is a statement of trust in the credibility of the information referred to. The more other users trust a particular source, and the more high-reputation documents refer to it, the higher its reputation. As such, the use of this measure is implicitly connected to a model of how users form links between documents. If this model actually applies, whether it is changing over time and to what extent the strategic behavior of users targeted at improving the ranking of their documents impacts this model is currently not clear. Today's search engine providers acknowledge these issues and try to address them by a continuous refinement of ranking methods. Reportedly, major search engine providers like GOOGLE update their ranking algorithms several hundred times a year, continuously responding to optimisation strategies of users. In addition, interventions in which certain privileged employees can manually degrade the algorithmically computed reputation of individual sources that are suspiciously highly ranked are being used to counter these strategies. These questions and current problems highlight that graph-based rankings cannot give a conclusive answer to the question of the reputation of a source of information, unless they are aligned with a substantiated model for the social processes that influence the formation of links.

In addition to the question what is the reputation of a source of information, significant additional efforts are currently being undertaken to improve the precision of search results based on a *profiling of users*. The interaction with any information system necessarily leaves digital traces that are increasingly being used to improve and personalise the ranking of information in future requests. To give a simple example, a user searching for the term "jaguar" and who exclusively clicks on results related to the animal, is likely to be interested in the animal also in future requests containing the term "jaguar". Even more, such information about the interest profile of a user can then be used to rank results of a new search query "beetle" in such a way, that information about insects are ranked higher than information about the car. These current developments highlight the fact that the relevance of information is not something that can be defined globally. Increasingly, the notion of relevance depends on who is searching for the information and it may even change as a user interacts with the system.

4.2.2 Social Information Filtering

In addition to the approaches discussed above, *explicit social mechanisms* are increasingly being introduced in a number of information systems. For this, information about the *social context* of users are typically being used as they have recently become available due to the widespread adoption of Online Social Networks. Here,it is assumed that users connected via a social tie are – at least in some respect – similar to each other, a sociological concept usually referred to as *homophily*. In addition to using the profile of a given user for the ranking of information, this concept suggests to additionally take into account the profile of their peers. Referring again to the previous example, a user who is connected to a number of other users that are interested in wild cats, may see corresponding results ranked higher when searching for the term "jaguar". Similarly, one can consider approaches where pieces of information that are – for some reason – popular among a large fraction of the population, generally ranked higher for other users as well, thus further enforcing their popularity.

Above, we have summarised current developments in the ranking of information based on the reputation of the source and the interests of users. We have further argued that the amount of information matching almost any possible search criterion is continuously rising. Hence, and due to the fact that users very rarely even look at those search results that are not ranked among the first, any *ranking* of information can effectively also be seen as a *filtering* of information. As a result, those documents consistently ranked last or those scientific publications published in the lowest-ranked journals will effectively be filtered out from the collective attention of users. Considering that much – if not most – of the information available in information systems like the World Wide Web is of interest only to a negligible fraction of users, one is tempted to question the model of *global* information spaces which – as they contain *all* information – require more and more sophisticated filtering techniques to extract those information relevant for a certain user or in a given context. One may propose instead a model consisting of numerous *local* spaces, across which only the most relevant and important information can propagate *globally*. It is essentially this model, which has recently resulted in *Online Social Networks* becoming increasingly popular as a general source of information, thus justifying to view them as *social information systems* [2, 6, 21, 22, 25]. Stories, pictures and news posted by users about their hobbies, interests or their life in general are generally of interest only to a local audience that reaches a few steps at most into the social network of a user. However, a user may still occasionally provide information that is of interest to a wider audience or, rarely, even to the whole population. Explicit propagation mechanisms like, e.g., a *Retweet* on

TWITTER or a *Like* on FACEBOOK allow users to propagate such information, thus allowing them to reach a larger audience. From the perspective of a prospective recipient of information, most of the information he encounters typically originates from within his closest social circles. At the same time, highly popular information originating from sources farther away is still likely to reach the user.

The appeal of this approach is the underlying idea that – by means of recommendations and explicit propagation mechanisms – the filtering of information is effectively left to the *collective intelligence* of users, rather than being computed centrally. Clearly, the arguments made above about graph-ranked methods apply here as well: Whether such a system works well and what exact notion of relevance applies to information that propagate far in the social network crucially depends on the social interactions between users. Both the evaluation and the design of such systems should thus be based on a sound model for the behavior of users. Even though certain advances have been made recently in the are of agent-based modelling, such models are not available for most of today's social information systems. Their initial design is thus mostly based on experience and *gut feeling* rather than scientific evidence, possibly fine-tuning interaction mechanisms based on observations made at run-time. In the offline world humans in a crowd often behave different from single individuals, thus giving rise to *emergent collective behavior* that is often difficult to anticipate or predict. Similar collective phenomena, and how they can be related to and influenced by individual behavior, need to be considered in social information systems. In today's social information systems, such phenomena frequently lead to surprising peaks of collective attention to particular users or pieces of information. They further influence how we consume information, how we perceive their importance and how we allocate our attention. Understanding the underlying mechanisms is hence of crucial importance.

4.3   Research Challenges

Based on the developments outlined in Sect. 4.2, in this section we will discuss a number of resulting challenges not only for computer science, but also for sociology, the study of complex systems, psychology and the philosophy of technology, thus justifying an interdisciplinary research agenda in social informatics.

4.3.1   Risks of Personalisation: The Filter Bubble

An important trend that can currently be observed in information systems is a shift in the notions of *relevance* that are being used for the ranking of information. Increasingly, the relevance of a piece of information is not defined in a global sense. The increasing availability of data on personal characteristics of users rather leads to a shift towards personalised notions of relevance, i.e. a piece of information may be relevant to one users, but less relevant to others. For this purpose, user profiles and social structures are being analysed and machine learning techniques are being applied and tuned to match the ranking and recommendation of information as closely to the users' preferences as possible. Product recommendation algorithms for example "learn" patterns in the buying behavior of customers in order to later recommend similar products in such a way that the probability of a purchase is maximised. Not only are these approaches based on the previous buying behavior of a particular customer, but also on the behavior of other customers with similar preferences.

As machine learning algorithms need data to be trained with, their dependency on prior user behavior is a concern particularly for the filtering and ranking of information. It has been argued that an increasing application of such filter methods will lead to – what has been termed – the *filter bubble* [28]. The filter bubble is an increasing limitation in the diversity of information users are exposed to, confining them mainly to those parts of the space of available information which, based on their previous behavior, matches their preferences. Once the amount of recommended information or products is large enough to completely consume the attention of a user, the chance of finding information that is not predictable from past behavior vanishes. A recent study of this potential problem [14] has revealed that the level of personalisation in the ranking of information in current major search engines is rather moderate. However, by means of an increasing integration of social features by major search engine providers, as well as the integration of data from large-scale social networks in novel search services, this problem is likely to aggravate in the future. Furthermore, through the spreading of information in social networks, even today users are consuming more and more information that has effectively been filtered by their social network [2, 6, 25] and which is thus less and less likely to extend and challenge their existing views. This development can not only change how individual users consume information, it can also affect the ability of forming substantiated opinions about issues of societal importance. It can further foster sociological and psychological effects like social reinforcement, group thinking, in-group biases and peer pressure. As such, the resulting limitation of information diversity can severely affect society. A research of such tendencies from the perspective of sociology and social psychology is utterly needed for an informed design of mechanisms in social information systems.

4.3.2 Discriminatory Effects of Information Filtering

Above, we have summarised risks associated with what one may call an "over-personalisation" of information systems, which can hinder a serendipitous discovery of information, and thus effectively shield off users from opposing perspectives. A related but different challenge is due to the fact that, apart from the actual *information content* produced by humans, a plethora of data on *personal characteristics of the producer* is becoming available in modern social information systems. Already today, data on users are becoming widely available that include their patterns of activity, their social and geographic context, age, gender, ethnics or even high-resolution mobility patterns. At the same time, data on the usage of information content by other users are increasingly being used to define quantitative proxies for the *relevance* – or even *quality* – of information. A natural result of this development is a recent stream of works investigating correlations between personal characteristics of users and the relevance and quality of information they provide. For new content provided by users, *Big Data* and *predictive analytics* techniques can then be combined to predict the potential quality or relevance of content, based on *who* has provided it. One may argue that any human assessment of the credibility of information is naturally taking into account its source, thus resulting in a prejudicial valuation which is also present in the offline world. On the one hand, prejudicial valuations which are based on prior experience with *particular individuals* can be attributed to the natural human tendency to build *trust* or *distrust* in their peers. On the other hand, prejudicial valuations which are merely based on personal characteristics like the age, gender, ethnics, geographical origin, or group membership of *unknown individuals* must be considered *discriminatory*. The ongoing push to a stronger filtering of information is likely to result in an increasing use of ranking methods that not only take into account *who* is the source of a piece of information, but also data on *personal characteristics*. This raises the important question which features of individuals can, or should, be used in the ranking of information, and under which conditions social information systems become discriminatory and thus unethical. Answering these questions of *machine or algorithm ethics* clearly goes beyond the capability of computer science, thus calling for a *social informatics* research agenda that integrates competence from the social sciences as well as from the philosophy of technology.

4.3.3   Effect of Information Filtering on Social Organisations

Increasingly available data on the social layer of social information systems can – and will – be used to refine information filtering mechanisms. Having discussed potential discriminatory effects that can negatively impact users of such systems, one can further consider potential effects of social information systems on *social organisations*. As an example, consider project management tools which are frequently used to foster collaboration and collectively process information in distributed software development projects. A particular type of such tools are issue tracking systems like e.g. BUGZILLA, which allow to submit and track information on software defects and to coordinate the efforts of software developers to fix them. Especially for software projects with a very large user base, it is common for software development teams to be overloaded with information that eventually turns out to be irrelevant for the processing of software defects [4, 40]. As a consequence, a number of works have studied how features of the actual information content can be used to filter and rank information in a way that mitigates overload and thus improves team efficiency [1,7]. In a recent study focusing on Open Source Software (OSS) projects, relations between the social embedding of users and the quality of information that they contribute have been investigated [40]. Findings suggest that a filtering of information which is merely based on the *social embedding* of users can outperform methods focusing on actual information content. Due to their reliance on part-time contributors with highly heterogenous capabilities and experience, the influence of user experience and reputation is likely to be particularly pronounced in OSS projects. As such, a ranking of information based on *experience, reputation and social embedding* of its source clearly bears a lot of potential. However, as argued above, it also entails discriminatory effects which are likely to particularly affect new and unexperienced members of an OSS community. A devaluation of information contributed by such new members can negatively affect their motivation and impair opportunities to learn, get feedback and gain experience. As such, introducing such ranking measures can have a devastating effect on the ability of a community to integrate and educate new contributors, thus impacting the structure and dynamics of their social organisation. Apart from this, admittedly extreme, scenario of how information systems may affect social organisation, more subtle examples exist in many of today's social information systems: Most systems which have explicit representations of a social network, not only prompt their users to provide information about their social relations. Based on methods which allow to predict future links in growing social networks [23], most of these systems actively *suggest or recommend* which social contacts users may be interested in. Hence, rather than merely monitoring the evolution of social structures, information systems have already started to actively shape the fabric of our society. What are the societal effects of this influence? Does it strengthen homophily? Does it affect group behavior or societal phenomena? Answering these questions is crucial for the *mechanism design* in future social information systems.

4.3.4   Epistemic Feedback in Social Information Systems

Summarising the challenges above, we are in a rather remarkable situation: information systems continuously collect large amounts of data on individual and collective human behavior, which – by means of large-scale statistical analyses – are used to tune predictive models that facilitate information ranking, personalisation and filtering. By this, information systems are likely to influence the very aspects of human behavior they are trying to model. Predictive models which are used to recommend links in online social networks are tuned using data on link formation processes, that are likely to change under the influence of link recommendation schemes. Similarly, ranking mechanisms naturally feed back on the social systems that collaboratively create information spaces. Network-based ranking schemes inevitably make – either implicit or explicit – assumptions about the *semantics* of a link: for citation-based measures of scholarly impact, a citation to a scientific article is typically interpreted as a statement of value that increases its reputation or impact.

Similarly, the number of hyperlinks pointing to a web site is thought to increase the reputation of a source. However, these links are not static. They are continuously created by humans, whose behavior is influenced by their ability to find information and their perceived importance. Being highly ranked can lead to an increase of citations to an article, or an increasing number of hyperlinks to a web page, thus reinforcing the ranking. However, the semantics of links that are the *reason for* an initial high ranking is likely to be quite different than of those links that *result from* a high ranking. As such, behavioral changes of humans in response to information ranking schemes can shift the semantics of links, invalidating ranking methods and requiring them to be updated and adjusted continuously. A related phenomenon can prominently be observed in the evolution of web search engines: By means of search engine optimisation strategies, commercial web site owners continuously strive to optimise their ranking, for instance by means of a strategic optimisation of hyperlinks. To discount for this behavior and to maintain a meaningful ranking, search engine providers continuously – and often secretly – update their ranking and filtering algorithms. It is tempting to interpret this complex coupling between social and technical systems as *epistemic feedback*, i.e. a feedback resulting from information systems that continuously measure and model social systems and social systems which respond to this modeling by means of a behavioral change. This feedback calls for a *systemic perspective* on the modeling and design of social information systems.

4.4   Towards a Systems Design of Social Information Systems

The challenges summarised in the previous sections highlight the need for a *systems approach* both to the design and quantitative analysis of social information systems. Such an approach necessarily integrates both the *social* and the *technical* layer of information systems and – as pointed out in the previous section – requires expertise beyond computer science. *Complex systems theory* is a particularly promising conceptual framework which – like for other types of *interwoven systems* [38] – can facilitate the *systems design* of social information systems. In the following, we specifically highlight three lines of research that are currently pursued in the study of complex systems and discuss their relevance in the context of information ranking.

4.4.1   Data-Driven Modeling of Social Systems

A crucial prerequisite for a reasonable design of socio-technical systems, is to improve our quantitative understanding of large-scale social systems. A promising development in this direction is the increasing adoption of data-driven modeling approaches, which are facilitated by the availability of large data sets on individual and collective human behavior. Not only does this approach allow to further test and develop sociological theories in ways that were unimaginable before. It also allows to improve our understanding how human behavior (i) influences and (ii) is being influenced by interactions mediated by information and communication technologies. To give an example, the use of modern text classification and statistical analysis techniques has recently allowed to model how emotional expressions influence the spreading of information, and thus their perceived importance [10, 31, 37]. It is interesting to study whether such models can foster our understanding of the *emotional dimension* of popularity, attention and relevance in social information systems, and whether they can be applied to improve information ranking and filtering schemes.

Another interesting research direction is the agent-based modeling of social behavior [35]. Models capturing the search and exploration behavior of users, and how it is influenced by ranking and filtering schemes, could possibly be used to study how users *collectively* consume information, thus addressing the challenges introduced in Sect. 4.3.1. Appropriate agent-based models can be used to study conditions for the emergence of detrimental collective behavior [24, 29, 30]. They can further be used to investigate possible countermeasure, like, e.g., a conscious and cautious introduction of *noise* in search results or recommendations. While such strategies may decrease the efficiency of information retrieval for *individual* users, they can possibly result in a better *collective* exploration of available information. Clearly, the potentially resulting transition from *individual* to *collective utility* of information systems introduces interesting further questions: What incentives do private providers of information systems have to introduce such mechanisms? Will it foster the perception of search engines and information systems as *public utility* of significant societal importance?

4.4.2 Complex Networks as Macroscopic Approach to Complex Systems

Apart from models which focus on the behavior of individual agents in social systems, another interesting, and highly active, line of research focuses on the *complex network topologies* formed by interacting humans or interlinked pieces of information. On the one hand, stochastic models for such complex networks provide a simple macroscopic approach to study basic topological features and hence better understand the characteristics of complex systems. On the other hand, network-analytic measures developed in the field of complex networks have proven useful to study particular empirical network topologies and answer important questions like, e.g., which users in a social network are most central or which piece of information in an information network is most relevant. Due to the availability of data both on information networks and social networks, this perspective is likely to play a key role in the definition of future ranking methods. Current relevant research addresses, for instance, novel network-based ranking algorithms, which – rather than computing a single, global ranking of nodes – provide a personalised ranking which can differ based on the perspective of a particular other node in the network [11, 15, 18, 39]. While most of these approaches are limited to *static networks*, time-varying network topologies, so-called *temporal networks*, have recently entered the focus of research [16]. Among those works, the question how the *order* of interactions in complex networks affects dynamical processes was recently addressed [32, 33, 36]. In the context of information systems, these works open new perspectives for the development of network-based ranking and clustering mechanisms that not only take into account the structure of complex networks, but also their temporal ordering [36].

Finally, considering possible cornerstones for a *social informatics* research agenda, a remarkable characteristic of the complex network perspective is that it provides a methodological framework that has long been used in sociological research. Studies of how humans route and retrieve information through their social circles used the network perspective, and have highlighted the importance of so-called *weak ties*, contacts *outside* the closest social circles of an individual (so-called *strong ties*). Sociological studies have shown that such weak ties are essential for the propagation of relevant information and for efficiently "navigating" social networks [12]. Linking *strong ties* to the filter bubble surrounding users in social information systems, it is tempting to relate this sociological finding, as well as its network interpretation, to the serendipity of information systems. Similar like the original study of weak ties, a detailed analysis of the importance of particular social structures for the spreading of information – and thus the filtering of information imposed by the social network – is important. In particular, it seems crucial for information systems to not undervalue information originating from sources with whom users seemingly have not much in common or with whom they maintain only loose contact. These may be exactly those weak ties through which – occasionally – the most valuable information enters a user's social circles, even though in most other cases information may be only of marginal relevance.

4.4.3   Multiplex Modeling of Socio-Technical Systems

Above, we have outlined that the theory of complex networks is being used both (i) to define relevance measures for networked information systems and (ii) to better understand the structure and dynamics of social systems and the flow of information there within. Developing into a kind of "lingua franca" that allows to model information systems and social systems in the same mathematical framework, the question arises whether network theory can be a used to analyze the combined system as well. Dealing with two networks, a network of interlinked information and a network of social relations, a so called *interdependent* network perspective seems to be appropriate for this purpose [9]. From a graph theoretic point of view, an interdependent network is a combination of two separate network layers, in which nodes are additionally connected by links that span across layers. Two prominent

a **b**

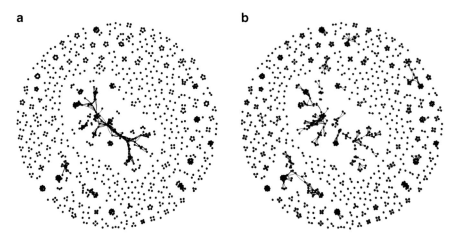

Fig. 4.1 Citation and collaboration network of authors at a computer science conference. (**a**) Citations between authors (**b**) Coauthorship between authors

examples are (i) hypertext systems like the World Wide Web (information layer) and Online Social Networks (the social layer) or (ii) citations between scientific papers (information layer) and collaboration networks of scientific coauthorships (social layer). An example for a multi-layer network of scientific collaborations and citations is shown in Fig. 4.1 for a major computer science conference. Here, nodes in each layer represent authors that either cite each other (a) or collaborate with each other on a paper (b). Identical nodes are placed at the same position in the two network layers. A comparison of the two network layers reveals a significant overlap between these structures. It is mainly authors that are close in the collaboration network, who cite each other and vice-versa. Especially in the case of scientific citations, which are often used to rank scientific articles, individual scientists or even institutions, the importance of taking a multi-layer perspective to these networks becomes apparent. Such a perspective has recently been used to provide interesting insights. In [34] the interlinked system of scientific publications and scientific coauthorships was considered. Based on the common assumption that citations are a direct measure of scientific success, this study was able to show that in a data set of more than 100,000 publications, citation success of publications can be predicted with high precision using only information on the social network generated by collaborating authors. This result provides evidence that in social information systems the social aspect can have considerable implications on the structure of the information network. Furthermore, the interlinked network perspective provides opportunities to define novel network-centric relevance measures which can possibly take into account both the social and the semantic layer of a system [13], thus allowing to define more reasonable ranking measures.

4.5 Conclusions

Looking back to the last century, it is clear that the history of computer science as a field of research is closely coupled with the development of information systems, and their widespread adoption in society. Over the decades, challenges involved in the contemporary use of such systems have significantly contributed to the development of database systems, formal languages, hardware technologies, computer networks, and machine learning to name only a few areas. One of the grand challenges involved in the design of future information systems is that they will be tightly interwoven with *social structures and phenomena*. As such, information systems will not only be influenced by the ways how humans and society use them, they also shape collective human behavior and thus societal developments. In this chapter we have argued that this development questions some of the approaches to information filtering and ranking that are being used in today's large-scale information systems. The increasing importance of their *social dimension* necessitates a dialogue between computer science and the social sciences, which is facilitated by recent advances in the data-driven modeling of social systems and their study from the perspective of (multi-layered) complex networks and complex systems. We are thus looking forward to an exciting era of interdisciplinary research on social information systems which, apart from being challenging scientifically, can have a significant beneficial impact on our society.

Acknowledgements IS and FS acknowledge support by the Swiss National Foundation, grant no. CR3I1_140644/1. We further acknowledge support by COST action TD1210: Analyzing the dynamics of information and knowledge landscapes.

References

1. Anvik J (2006) Automating bug report assignment. In: Proceedings of the 28th international conference on software engineering, ICSE '06, Shanghai. ACM, New York, pp 937–940. 10.1145/1134285.1134457. http://doi.acm.org/10.1145/1134285.1134457
2. Bakshy E, Rosenn I, Marlow C, Adamic L (2012) The role of social networks in information diffusion. In: Proceedings of the 21st international conference on world wide web, Lyon. ACM, pp 519–528
3. Berners-Lee T, Groff JF (1992) Www. SIGBIO Newsl 12(3):37–40. 10.1145/147126.147133. http://doi.acm.org/10.1145/147126.147133
4. Bettenburg N, Just S, Schröter A, Weiss C, Premraj R, Zimmermann T (2008) What makes a good bug report? In: Proceedings of the 16th ACM SIGSOFT international symposium on foundations of software engineering, SIGSOFT '08/FSE-16, Atlanta. ACM, New York, pp 308–318. 10.1145/1453101.1453146. http://doi.acm.org/10.1145/1453101.1453146
5. Brin S, Page L (1998) The anatomy of a large-scale hypertextual web search engine. Comput Netw ISDN Syst 30(1–7):107–117. 10.1016/S0169-7552(98)00110-X. http://linkinghub.elsevier.com/retrieve/pii/S016975529800110X
6. Bughin J, Corb L, Manyika J, Nottebohm O, Chui M, de Muller Barbat B, Said R (2011) The impact of internet technologies: search. McKinsey&Company, High Tech Practice

7. Cubranic D, Murphy GC (2004) Automatic bug triage using text categorization. In: Maurer F, Ruhe G (eds) SEKE, Banff, pp 92–97

8. Frisse M (1988) Searching for information in a hypertext medical handbook. Commun ACM 31(7):880–886

9. Gao J, Buldyrev SV, Stanley HE, Havlin S (2012) Networks formed from interdependent networks. Nat Phys 8(1):40–48

10. Garcia D, Schweitzer F (2012) Modeling online collective emotions. In: Proceedings of the 2012 workshop on data-driven user behavioral modelling and mining from social media-DUBMMSM '12, CIKM2012, Maui. ACM, New York, p 37. 10.1145/2390131.2390147. http://dl.acm.org/citation.cfm?id=2390131.2390147

11. Goldman R, Shivakumar N, Venkatasubramanian S, Garcia-Molina H (1998) Proximity search in databases. In: Proceedings of the 24th international conference on very large data bases, VLDB '98, New York. Morgan Kaufmann Publishers, San Francisco, pp 26–37. http://dl.acm.org/citation.cfm?id=645924.671346

12. Granovetter M (1973) The strength of weak ties. Am J Sociol 78(6):1

13. Halu A, Mondragón RJ, Panzarasa P, Bianconi G (2013) Multiplex pagerank. PLoS ONE 8(10):e78,293. 10.1371/journal.pone.0078293. http://dx.doi.org/10.1371%2Fjournal.pone.0078293

14. Hannak A, Sapiezynski P, Molavi Kakhki A, Krishnamurthy B, Lazer D, Mislove A, Wilson C (2013) Measuring personalization of web search. In: Proceedings of the 22nd international conference on world wide web, Rio de Janeiro. International World Wide Web Conferences Steering Committee, pp 527–538

15. Haveliwala T, Kamvar S, Jeh G (2003) An analytical comparison of approaches to personalizing pagerank. Technical report 2003–35, Stanford InfoLab. http://ilpubs.stanford.edu:8090/596/

16. Holme P, Saramäki J (2012) Temporal networks. Phys Rep 519(3):97–125

17. Introna L, Nissenbaum H (2000) The politics of search engines. IEEE Comput 54–62. http://ieeexplore.ieee.org/stamp/stamp.jsp?tp=&arnumber=816269

18. Jeh G, Widom J (2003) Scaling personalized web search. In: Proceedings of the 12th international conference on world wide web, WWW '03, Budapest. ACM, New York, pp 271–279. 10.1145/775152.775191. http://doi.acm.org/10.1145/775152.775191

19. Kleinberg J (1999) Authoritative sources in a hyperlinked environment. J ACM (JACM) 46:604–632. http://dl.acm.org/citation.cfm?id=324140

20. Kraemer F, Overveld K, Peterson M (2010) Is there an ethics of algorithms? Ethics Inf Technol 13(3):251–260. 10.1007/s10676-010-9233-7. http://link.springer.com/10.1007/s10676-010-9233-7

21. Kwak H, Lee C, Park H, Moon S (2010) What is Twitter, a social network or a news media? In: Proceedings of the 19th international conference on world wide web, Raleigh. ACM, pp 591–600

22. Lerman K, Ghosh R (2010) Information contagion: an empirical study of the spread of news on Digg and Twitter social networks. ICWSM 10:90–97

23. Liben-Nowell D, Kleinberg J (2003) The link prediction problem for social networks. In: Proceedings of the twelfth international conference on information and knowledge management, CIKM '03, New Orleans. ACM, New York, pp 556–559. 10.1145/956863.956972. http://doi.acm.org/10.1145/956863.956972

24. Mavrodiev P, Tessone CJ, Schweitzer F (2012) Effects of social influence on the wisdom of crowds. In: Proceedings of Collective Intelligence 2012, CoRR abs/1204.3463. http://arxiv.org/html/1204.2991

25. Naaman M, Boase J, Lai, CH (2010) Is it really about me?: message content in social awareness streams. In: Proceedings of the 2010 ACM conference on computer supported cooperative work, Savannah. ACM, pp 189–192

26. Newman MEJ (2010) Networks: an introduction. Oxford University Press, Oxford/New York. 10.1093/acprof:oso/9780199206650.001.0001. http://www.oxfordscholarship.com/view/10.1093/acprof:oso/9780199206650.001.0001/acprof-9780199206650
27. Page L, Brin S, Motwani R, Winograd T (1999) The PageRank citation ranking: bringing order to the web. Technical report, Stanford InfoLab. http://ilpubs.stanford.edu:8090/422
28. Pariser E (2011) The filter bubble: what the Internet is hiding from you. Penguin, New York
29. Parunak HVD (2011) Swarming on symbolic structures: guiding self-organizing search with domain knowledge. In: Eighth international conference on information technology: new generations, ITNG 2011, Las Vegas, 11–13 Apr 2011, pp 896–901
30. Parunak HVD, Downs E, Yinger A (2011) Socially-constrained exogenously-driven opinion dynamics: explorations with a multi-agent model. In: SASO, Ann Arbor. IEEE, pp 158–167
31. Pfitzner R, Garas A (2012) Emotional divergence influences information spreading in Twitter. In: Sixth International AAAI Conference on Weblogs and Social Media, AAAI ICWSM 2012, Dublin, pp 2–5. http://www.aaai.org/ocs/index.php/ICWSM/ICWSM12/schedConf/presentations
32. Pfitzner R, Scholtes I, Garas A, Tessone CJ, Schweitzer F (2013) Betweenness preference: quantifying correlations in the topological dynamics of temporal networks. Phys Rev Lett 110(19):198,701. 10.1103/PhysRevLett.110.198701. http://prl.aps.org/abstract/PRL/v110/i19/e198701
33. Rosvall M, Esquivel AV, Lancichinetti A, West JD, Lambiotte R (2013) Networks with memory. arXiv preprint, arXiv:1305.4807
34. Sarigöl E, Pfitzner R, Scholtes I, Garas A, Schweitzer F (2014) Predicting scientific success based on coauthorship networks. Working paper, to appear in EPJ Data Science
35. Scheitzer F (2003) Brownian agents and active particles. Collective dynamics in the natural and social sciences. Springer series in synergetics. Springer, Berlin
36. Scholtes I, Wider N, Pfitzner R, Garas A, Tessone CJ, Schweitzer F (2013) Slow-down vs. speed-up of diffusion in non-markovian temporal networks. arXiv preprint. http://arxiv.org/abs/1307.4030
37. Šuvakov M, Garcia D, Schweitzer F, Tadić B (2012) Agent-based simulations of emotion spreading in online social networks. ArXiv e-prints. http://arxiv.org/abs/1205.6278
38. Tomforde S, Hähner J, Seebach H, Reif W, Sick B, Wacker A, Scholtes I (2014) Engineering and mastering interwoven systems. In: Proceedings of the 2nd international workshop on self-optimisation in organic and autonomic computing systems (SAOS 2014), Lübeck
39. Walter FE, Battiston S, Schweitzer F (2009) Personalised and dynamic trust in social networks. In: Proceedings of the third ACM conference on Recommender systems – RecSys '09, New York. ACM, New York, pp 197–204. 10.1145/1639714.1639747
40. Zanetti MS, Scholtes I, Tessone CJ, Schweitzer F (2013) Categorizing bugs with social networks: a case study on four open source software communities. In: Proceedings of the 35th international conference on software engineering, ICSE '13, San Francisco, 18–26 May 2013, pp 1032–1041
41. Zwass V (2014) In: Encyclopaedia britannica http://www.britannica.com/EBchecked/topic/287895/information-system. Retrieved 11 Feb 2014

Chapter 5
Using Weighted Interaction Metrics for Link Prediction in a Large Online Social Network

Oliver Posegga, Kai Fischbach, and Martin Donath

Abstract There has been a considerable amount of recent research on the link prediction problem, that is, the problem of accurately predicting edges that will be established between actors in a social network in a future period. With the cooperation of the provider of a German social network site (SNS), we aim to contribute to this line of research by analyzing the link formation and interaction patterns of approximately 9.38 million members of one of the largest German online social networks (OSN). It is our goal to explore the value of users' interaction frequencies for link prediction based on metrics of local structural similarity. Analyzing a random sample of the network, we found that only a portion of the network is responsible for most of the activity observed: 42.64 % of the network's population account for all observed interactions and 25.33 % are responsible for all private communication. We have also established that the degree of recent interaction is positively correlated with imminent link formation – users with high interaction frequencies are more likely to establish new friendships. The evaluation of our link prediction approach yields results that are consistent with comparable studies. Traditional metrics seem to outperform weighted metrics that account for interaction frequencies. We conclude that while weighted metrics tend to predict strong ties, users of SNS establish both strong and weak ties. Our findings indicate that members of an SNS prefer quantity over quality in terms of establishing new connections. In our case, this causes the simplest metrics to perform best.

O. Posegga (✉) • K. Fischbach
Department of Information Systems and Social Networks, University of Bamberg,
An der Weberei 5, 96052, Bamberg, Germany
e-mail: oliver.posegga@uni-bamberg.de; kai.fischbach@uni-bamberg.de

M. Donath
voola GmbH, Brüsseler Platz 8, 50672, Köln, Germany
e-mail: martin.donath@voola.de

© Springer International Publishing Switzerland 2014
K. Zweig et al. (eds.), *Socioinformatics - The Social Impact of Interactions between Humans and IT*, Springer Proceedings in Complexity,
DOI 10.1007/978-3-319-09378-9_5

5.1 Introduction

A common way of analyzing complex systems is to model them as networks. Entities of a system can be represented as nodes and their relationships can be converted to edges. For example, the nature of scientific collaborations can be studied by mapping authors to nodes and their collaborations to edges. The topology of the resulting graph can then be used to gain insight into the underlying system's mechanics. This method has been used for a long time across several scientific disciplines, including biology, physics, economy, computer science, and sociology. The popularity and success of this approach eventually led to the emergence of a new scientific discipline, network science, which is dedicated to the development and improvement of methods and techniques to analyze and understand networks in general [5, 26]. Even though networks in different scientific domains can exhibit fundamental differences in their structural properties [22], the growing number of well-defined tasks, methods, and problems related to their general analysis is adaptable for many of them.

One of these problems has received a considerable amount of attention: namely, the *link prediction problem*. In general, it can be described as the problem of predicting the likelihood of occurrence for all as yet nonexistent or unknown edges in a network. For social networks, the problem has been defined by Liben-Nowell and Kleinberg as follows: "Given a snapshot of a social network at time t, we seek to accurately predict the edges that will be added to the network during the interval from time t to a given future time t'." [15].

This problem is relevant to many scientific communities [17], but it has become of particular interest to those focusing on networked systems comprising social actors. Finding good solutions to the problem requires a proper understanding of the process of *link formation*, which reflects the decision of individuals to engage in social interactions. Considerable research has been dedicated to this straightforward yet difficult to understand process and its effect on overall network properties. Some scholars have taken a macroscopic approach and analyzed longitudinal datasets, seeking to shed light on the development of global network properties over time (e.g. [13, 14]). Others have been microscopic, striving to identify patterns of link formation [12, 2, 14, 23]. It is as part of these efforts that Liben-Nowell and Kleinberg defined the link-prediction problem. In their analysis of co-authorship networks, they gained insight into the effect of structural network properties on the likelihood of future link formation. Motivated by their work and the growing need to understand the mechanisms behind the evolution of networks, a large variety of approaches to this problem has been developed [29, 17].

The interest in the topic is reinforced by the constantly growing number of *social network sites* (SNS) and large *online social networks* (OSN). Boyd and Ellison [6] account for the launch of 11 major SNS from 1997 to 2002 (e.g., LiveJournal, Friendster) and 29 launches (e.g., Facebook, Twitter, and YouTube) between 2003 and 2006, a trend that indicates the increasing popularity and cultural relevance of such sites. According to their definition, a SNS is a web-based service that allows individuals to construct a profile including a visible network of connections to others. While many of these SNS provide different features and target different social groups, they all share a common goal: growth. Once they reach a critical mass, network effects facilitate further growth.

While such effects are of major importance to SNS, they come with downsides. First, a rapidly growing number of community members require a scalable and reliable technological infrastructure, which involves high costs. Second, and even more important for our work, it strongly requires mechanisms to make the platform easy to use for its members. One important aspect of a social network site's ease of use is its ability to reduce search costs and increase transparency for its users by enabling them to find new friends easily (i.e., people with whom they wish to connect). A common way to achieve this is to recommend an individual list of likely connection candidates to each user. Such a list's value to a user depends on the relevance of the recommended candidates. If it contains a high number of relevant suggestions, users are likely to establish connections with some of them and thereby contribute to the network's internal growth. The problem of making relevant suggestions to a user can be referred to as the *link recommendation problem* [3, 24, 25], which is a closely related to the link prediction problem. Hence, accurate and scalable link prediction algorithms can provide a direct value to SNS operator, by making their platforms easier to use and more valuable for members.

In short, finding good solutions for the link prediction problem in social networks is important from a theoretical and practical perspective. At the same time, finding good solutions is not easy – especially in large social networks, where we typically observe an odd ratio of the average number of friends per user to the total number of members in the network (i.e., large social networks are sparse). Moreover, the phenomenon of social interaction is highly complex. As Watts [27] described, it is the result of the history of interactions between a large numbers of heterogeneous individuals, their behavior, and their interdependent decisions; understanding them by focusing exclusively on structural network properties is difficult.

In this work, we focus on improving local similarity metrics used in neighbor-based link prediction approaches that have been discussed by Liben-Nowell and Kleinberg [15] and Lü and Zhou [17]. We argue that traditional versions of such metrics are designed only for unweighted networks, and hence are usually applied to simple social graphs, ignoring a large portion of additional network data generated by SNS. Accordingly, we modify traditional metrics and apply them to a social network site's interaction graph [28], which is generated from direct user interactions. In addition, we take weighted variations of this graph into account and apply modified versions of traditional metrics to those variations. We conduct an empirical experiment in cooperation with the operator of a large German SNS to test the performance of this approach. Our findings are consistent with the results of related studies. The traditional metrics seem to dominate over the more complex variations we used. When applied to a weighted combination of the social and interaction graph, the metrics Adamic/Adar [1] and Resource Allocation [17] seem to perform best, followed closely by the metric Common Neighborhood. In alignment with Lü and Zhou [16], we draw the conclusion that weighted metrics tend to predict strong ties, whereas users of SNS establish both strong and weak ties, as described by Donath and Boyd [7].

The remainder of this manuscript is organized as follows. In Sect. 5.2, we provide background information on similarity-based link prediction and present the metrics used for our experiment. Section 5.3 is dedicated to introducing the setup and methodology of our experiment. In Sect. 5.4, we present our results, beginning with a brief overview of the dataset's general properties, followed by a performance evaluation of the metrics. In Sect. 5.5, we discuss the results and reflect on them in the context of our initial theoretical considerations. We discuss potential limitations in Sect. 5.6 and, finally, draw our conclusions in Sect. 5.7.

5.2   Theoretical Background and Related Literature

As described above, the focus of this work is on metrics used for link prediction in large online social networks based on SNS. Such networks differ from networks based on other platforms, such as message boards or e-mail systems. In those networks, there is typically only one type of user interaction that is translated into edges of a corresponding network (e.g., mutual engagements in a board conversation or e-mail communication). In contrast, SNS offer their members more than one way of interaction. Actors typically have access to a variety of interaction-enabling features: they can befriend each other, communicate privately and publicly, update their status messages, interact in and with groups, attend events, and make use of many more features with which to engage in social interaction. Accordingly, one could argue that within a SNS there is, in fact, more than one network to analyze since each feature could be understood as a source for a single network with its own characteristics. Taking a closer look at data collected from Facebook, Wilson et al. [28] address this topic. While they are primarily interested in understanding the real-world relevance of SNS-based interaction, they introduce the concept of the *interaction graph* (*IG*) and compare it to the *social graph* (*SG*).

A social network site's *social graph* is generated from mutual connections between pairs of actors (often referred to as friendships). Edges of this graph are naturally unweighted, undirected, and persistent, that is, they are not subject to any kind of natural decay. As Donath and Boyd [7] describe, connections represented by edges in this graph are not necessarily related to a mutual focus or interest of the actors involved. According to them, the overall costs of creating this type of edge are low and, hence, people tend to add many *friends* that include even those with whom they do not share a deep relationship. This is the primary reason Wilson et al. [28] look for better indicators of real-world relationships in SNS and ONS.

In doing so, Wilson et al. [28] introduce the *interaction graph*, which is generated from interaction data (e.g., private or public communication). Following their definition, it is a subgraph of the social graph that contain the same set of nodes, but only those edges that have been subjected to a minimum number of n interactions during a time interval $[t, t']$. Hence, in contrast to the social graph, edges of the interaction graph are subject to decay. By definition, this graph is unweighted and undirected, but it can be adjusted to account for both edge weights and directions. Wilson *et al.* find significant differences in the structural properties of the social and interaction graphs. Their findings suggest that the total number of friends with whom one genuinely interacts seems to be limited.

This finding indicates that the costs of adding and maintaining connections in the interaction graph are higher than in the social graph – an observation consistent with the findings of Hill and Dunbar [11], who report that people are limited in their ability to maintain a large number of active relationships. Other studies show similar results (e.g., [23, 14]). Another central finding of the work of Wilson et al. is that the interaction graph seems to provide richer information about social relationships between network members than the social graph. This finding is reinforced by Gilbert and Karahalios [9], who find that communication-based events (i.e., interaction graph-based information) are better predictors of strong ties [10] than traditional structural features (i.e., social graph-based information).

Summarizing the above, many studies conclude that users' interaction patterns provide rich insights into their relationships. Such insights differ from those gained by analyzing the social graph. We adapt the distinction between the *SG* and the *IG* to explore the quality of link prediction based on the *IG* and combinations of both graphs, since simple approaches to link prediction, especially structural similarity-based algorithms, usually focus exclusively on the *SG*.

Further, we extend the concept of the *IG* by considering its weighted version, which we refer to as the *weighted interaction graph (WIG)*. Like the *IG*, it is a subgraph of the *SG*. In contrast to the *IG*, the *WIG*'s edges are weighted with all interactions between the corresponding actors during the time interval under consideration. Accordingly, the *WIG* carries more information than the original *IG*, making it more interesting for our link prediction approach. However, the *IG* and *WIG* completely neglect the information carried by the *SG*. Hence, we also experiment with a simple combination of the *SG* and the *WIG*, that is, a *combined graph (CG)*. In a first approach, we construct the *CG* by initializing it with the *SG*, where we consider each edge to be weighted with a weight of 1. Subsequently, we merge this graph with the corresponding *WIG* by adding the edge weights of edges that are present in both graphs.

Before we present our experimental setup to evaluate the performance of our approach, we provide background information on similarity-based link prediction and the corresponding metrics.

In our approach, we focus on link prediction using structural similarity-based algorithms [17, 15] and neighbor-based similarity metrics, largely because of their high computational efficiency and the low amount of information they require to make valuable predictions. More important, this approach gives us full control over the amount of information used for the prediction, thus allowing us to compare the quality of predictions derived from the *SG* with predictions derived from the *IG* and their combination.

Similarity-based algorithms follow the basic assumption that two nodes of a network are more likely to establish a future connection if they are close or, in terms of their structural properties, similar to each other. Such similarity, or proximity, can be quantified using proximity metrics, which can be utilized to assign a score to each pair of unconnected nodes in a network. Pairs of nodes can be sorted and ranked based on such scores. Furthermore, the top-k (where k is typically a number between 10 and 100) ranked pairs can be classified as the most similar actors, which are assumed to be most likely to share a future connection and, hence, are predicted to be future acquaintances.

The *proximity metrics* used by similarity-based algorithms can be categorized by the type of information they require. One category of metrics focuses on nodal attributes (e.g., demographic information, such as gender, age, or geographic location). Other so-called *structural similarity metrics* focus exclusively on topological information. Such metrics have been subject to the fundamental work of Liben-Nowell and Kleinberg [15], who further divided them into *path-* and *neighbor-based* metrics. Liben-Nowell and Kleinberg tested several of these metrics on scientific collaboration networks and compared their performance to random predictions, concluding that even simple structural similarity metrics are capable of producing valuable predictions. This, it turns out, is particularly true for the neighbor-based metrics, which are the simplest metrics considered.

Neighbor-based metrics exhibit a limited perspective on the network. When used for link prediction, such metrics neglect pairs of actors separated by a path with a length above 2. In other words, they consider only pairs of actors with at least one mutual friend. The set of mutual friends is also referred to as neighborhood. While the restriction to this minimal distance seems to be strong, many studies show that a major portion of newly established links connects actors who have been sharing at least one mutual friend [14]. In fact, the phenomenon behind this observation, called *triadic closure*, is well known and has been studied in many networks [12, 21].

To allow for a more precise presentation of the neighbor-based metrics discussed in this work, we first introduce a mathematical notation consistent with the one used by Liben-Nowell and Kleinberg [15]. Given a set of nodes V and a set of edges E, we denote the corresponding graph by $G(V, E)$. Unless stated otherwise, we assume that $G(V, E)$ is unweighted and undirected. Since we are working with social networks, nodes are also referred to as actors or friends and, depending on the context, edges are referred to as friendships or connections. Given a node $v \in V$, we formulate the set of its neighbors (i.e., all nodes directly connected to v) as $\Gamma(v)$. Hence, the degree of a node can be written as $|\Gamma(v)|$. We proceed with the introduction of five well-known neighbor-based metrics, which we use as baseline metrics for our experiment.

Common Neighborhood (*CN*) is the most basic of the neighbor-based metrics. Given two nodes $x, y \in V$, it describes their similarity in terms of mutual neighbors. Using this metric for link prediction in social networks reflects the assumption that two actors are more likely to become friends when they already have a high number of mutual acquaintances [20]. The corresponding formula for *CN* is:

$$CN = |\Gamma(x) \cap \Gamma(y)|$$

Jaccard's coefficient (JC), a similar metric, resembles a normalized version of *CN*. The number of friends shared by two actors is divided by the number of friends connected to one or the other actor:

$$JC = \frac{|\Gamma(x) \cap \Gamma(y)|}{|\Gamma(x) \cup \Gamma(y)|}$$

Preferential Attachment, another well-studied metric, is a fundamental component of several models of network evolution [4]. It has been subject to the studies of Newman [19] and inhibits the assumption that nodes with a many connections have a higher likelihood of being involved in future link formation. It is defined as:

$$PA = |\Gamma(x)| \times |\Gamma(y)|$$

Adamic/Adar (AA) is a similarity metric that aggregates the number of shared features between two nodes of a network [1]. In addition, it considers rare features to be more valuable. In terms of social networks, this means that the metric diminishes the value of mutual friends with a many connections and considers those with a few connections to be more valuable:

$$AA = \sum_{z \in \Gamma(x) \cap \Gamma(y)} \frac{1}{\log(|\Gamma(z)|)}$$

Resource allocation, a very similar yet differently motivated metric, was introduced by Zhou et al. [30]. It values rare common neighbors even more than *AA* and performs well; depending on the type of network, it can exceed the performance of *CN* and *AA* [17]. They defined it as follows:

$$RA = \sum_{z \in \Gamma(x) \cap \Gamma(y)} \frac{1}{|\Gamma(z)|}$$

The application of the CN, JC, PA, and AA metrics has been studied extensively on social networks. We use them as a baseline to benchmark the performance of our approach. We also consider the relatively newer *RA*, since it has performed well on a variety of networks. However, to the best of our knowledge, it has not yet been tested on a large online social network comparable to the one we use for our experiment.

Some of these metrics have been adjusted to work on weighted graphs to improve their performance by capturing multiple edges between nodes of a network [18, 16]. Murata and Moriyasu tested weighted versions of *CN*, *AA*, and *PA* on the network of the Japanese *Yahoo! Answers* (a Question-Answering Bulletin Board (QABB)) and showed that weighted metrics outperform their unweighted versions. Based on this work, Lü and Zhou [16] tested weighted versions of *CN*, *AA*, and *RA* on the USAir transportation network, the neural network of the worm C. elegans, and the CGScience co-authorship network. In contrast, they found an inferior prediction performance for the weighted versions of the considered metrics in all networks except for the neural network.

We adapt some of the weighted metrics proposed by Lü and Zhou [16] for our own experiment. In addition, we consider a weighted version of *PA*, which has been introduced by Murata and Moriyasu [18]. Finally, we present a weighted version of *JC*, which we developed for this experiment.[1] The metrics are represented by the following mathematical formulations, where x, y, and z are nodes of an undirected, weighted graph with edge weights $w(x, y)$ and $s(x) = \sum_{z \in \Gamma(x) \cap \Gamma(y)} w(x, z)$ denotes the weighted degree of node x:

$$WCN = \sum_{z \in \Gamma(x) \cap \Gamma(y)} w(x, z) + w(z, y)$$

$$WAA = \sum_{z \in \Gamma(x) \cap \Gamma(y)} \frac{w(x, z) + w(z, y)}{\log(1 + s(z))}$$

$$WRA = \sum_{z \in \Gamma(x) \cap \Gamma(y)} \frac{w(x, z) + w(z, y)}{s(z)}$$

$$WPA = s(x) \times s(y)$$

$$WJC = \frac{\sum_{z \in \Gamma(x) \cap \Gamma(y)} w(x, z) + w(z, y)}{s(x) + s(y)}$$

In this section, we introduced five well-known neighbor-based metrics, discussed their origins, and explained their weighted variants. While the studies on weighted metrics we reviewed try to improve prediction performance on several types of networks, they pay little attention to the underlying network specifics. In the following section, we describe the application of the weighted metrics to a set of graphs that accounts for actors' interaction frequencies in terms of private communication, and provide overall details on experimental setup and methodology.

---

[1] While we stick to the notation *WJC* because of the analogy to *JC*, we note that the *JC* metric has a more complex background. This version does not necessarily comply with its initial intention. We are interested primarily in the analogy of its interpretation, and hence this is merely a formal issue and of no further relevance for our work.

5.3 Methodology

In cooperation with a large German SNS operator, we evaluate the performance of the metrics discussed above on the different graphs (i.e., the *SG*, *IG*, *WIG*, and *CG*) in an experimental setup similar to that designed by Liben-Nowell and Kleinberg [15]. We were granted access to a snapshot of the platform's complete *SG*, taken on June 8, 2011. Moreover, we were allowed to track 49 types of events triggered by user interactions (e.g., writing a message, establishing a contact) for a 60-day period between June 8 and August 6, 2011. To perform our experiment, we split this timeframe into two periods of equal length, *training* and *test* (see Table 5.1). We use the data from the training period to predict future link formations for a randomly chosen subset of actors in the test period.

Therefore, we use the logged interaction events, including the creation and deletion of edges in the *SG*, to update the initial snapshot of the *SG* at t_0 to the end of the training period t'_0. Subsequently, we create the *IG*, *WIG* and *CG* as described in the previous section and set the parameter t to t_0 and n to 1. Accordingly, all three graphs are generated from 30 days of interaction. As mentioned earlier, a SNS provides a large variety of features from which one could derive an interaction graph. In our scenario, we focus on a very common SNS feature to derive our graphs, that is, *private communication*.[2] Consequently, two actors share an edge in the *IG* when they have established a mutual friendship on the SNS and communicated at least once ($n = 1$) during the training period ($t = t_0$). Moreover, the weight of an edge in the *WIG* reflects the sum of messages exchanged between two actors during the respective interval. The *CG* comprises nodes and aggregated edges from the *SG* and *WIG*.

To make our predictions for the test period, we pick a random sample S of 15 users from all users that have been active members of the SNS throughout the period of observation. More precisely, we draw our sample from all actors with at least five friends at the end of the training period and who establish at least five additional friendship connections during the test period.

We now ask the following question: Given an actor $s \in S$, who are the k most likely candidates C_s to establish a connection with s in the *SG* during the test period? Like Backstrom and Leskovec [3], we define the set of candidates as $C_s = D_s \cap L_s$,

Table 5.1 Intervals

Interval	Parameter	Date
Training	t_0	June 8th , 2011, 00:00:00 CEST
	t'_0	July 7th , 2011, 23:59:59 CEST
Test	t_1	July 8th , 2011, 00:00:00 CEST
	t'_1	August 6th , 2011, 23:59:59 CEST

---

[2]The SNS we analyzed enables private communication between users by providing an integrated direct messaging system through which users exchange text messages via a web-based interface.

where D_s refers to the set of *destination nodes* to which s will establish a connection in the test period; L_s denotes the set of *no-link nodes* an actor will not befriend in the test period. In other words, our goal is to pick k actors from C_s, capturing as many actors from D_s as possible.

To reduce the computational complexity, we restrict each set C_s to actors who share at least three common neighbors with s. Since actors with a low number of common neighbors are unlikely to establish a future link with s, we do not expect a significant loss in prediction accuracy due to this restriction [3].

After identifying all candidate nodes, we compute *CN*, *JC*, *AA*, *RA*, and *PA* for each pair of users (s, c) with $c \in C_s$ on the *SG* and *IG*. Similarly, we proceed for the weighted versions *WCN*, *WJC*, *WAA*, *WRA*, and *WPI* on the *WIG* and *CG*. Whenever we apply a metric M to a graph G, we refer further to the result as M_G. For example, CN_{SG} refers to the metric Common Neighborhood when it is computed based on the social graph.

As a result of this *scoring* procedure, we receive one score for each pair (s, c) for each metric and the respective graph to which it has been applied. In the next step, we *sort* each list in a descending order and *rank* the pairs of actors accordingly. For each user s, we now have 20 sorted and ranked lists of candidates – one for each computed metric.

Finally, we define the top k entries of each list to be our predictions for the test period. We discuss the results for $k \in \{100, 50, 20\}$ based on the following metrics.

To compare the prediction performance of the metrics considered, we need a method to evaluate the performance for each metric. Such a metric has to deal with two problems. First, we must expect a skewed distribution of negative and positive predictions – the applied algorithms are likely to create considerably more negative than positive predictions. In addition, the number of link formations observed during the test period will probably be much smaller than the number of link formation candidates. Second, each of the sample users considered creates an individual number of friendships during the test period. To evaluate the metrics across users, a suitable performance measure has to be robust with respect to the different number of positive and negative observations for each sample user.

The *area under the curve* (AUC) metric fulfills the criteria: it is a performance indicator typically used to evaluate prediction results. It can be derived from *receiver operating characteristics* (ROC) curve. Such a curve is a visualization of a metric's true-positive and false-positive rates. Accordingly, it is independent from the number of observed link formations and candidates per user.[3]

Based on the results of our experiment, we calculate the true-positive and false-positive rates for all of the 20 top k predictions for each of the sample users S. Next, we use the corresponding ROC curves to compute the individual AUC values and calculate average AUC per metric for each value of k.

To evaluate a metric's overall performance, we check whether a metric is able to achieve an average AUC greater than 0.5. If so, that metric is capable of scoring

---

[3]For more information on the topic of ROC analysis, we refer to Fawcett [8] and Lü and Zhou [17].

better than a random predictor. Accordingly, we conduct a one sample t-test for each metric, testing it against a value of 0.5. We use a confidence level of 95 % and report the resulting significance.

5.4 Results

This two-part section presents our results. In the first part, we provide descriptive statistics. In the second part, we give a full presentation of our experiment's results.

The initial SG comprises 513,419,650 edges between 7.4 million unique actors. In total, the graph contains 9.38 million actors. It is worth noting that the data suggest that 21.11 % of all known actors have 0 friends at t_0. Furthermore, the average platform user has approximately 110 friends. Users are limited to a maximum of 4,000 friends.

During the entire observation period, we logged 521,583,014 interactions. Within the 60 days of observation, 4.0 million users took part in at least one interaction and 2.7 million of those sent at least one private message. Thus, 42.64 % of all users were responsible for all interactions observed in 60 days and 25.33 % produced all outgoing private messages during that period.

Among the logged interactions, 11,690,430 friendships were established and 2,789,371 friendships were deleted. Some 87 % of all created friendships were formed between users who already existed in the initial snapshot of the SG. We took a random sample of 1,000 recently established friendships from the test period and found that 84 % of all new friendships are formed between actors with one or more common neighbors. Moreover, 80 % of all those new friendships are established between actors with a minimum of three common neighbors. Since we required all candidates C_s to have at least three common neighbors with a sample user s, this implies that we neglect approximately 20 % of all link formations that can be predicted using neighbor-based metrics. Additionally, we used the sample above to compute Spearman's rank correlation coefficient for the number of private messages sent by a user during the training period and the number of friendships established by the same user during the test period. We find a significant correlation of 0.528 between both variables, indicating that private communication frequencies are capable of predicting future link formations.

The users in our sample S established 241 friendships among 71,715 candidates C. Accordingly, they befriended 0.336 % of all considered candidates within the 30-day test period.

In Table 5.2, we present the average AUC values for all CN metrics and their top 100 predictions. We skip reporting the full results for the top 50 and top 20 predictions because they are similar to the results reported below and, in most cases, were not significant.

We find that CN_{SG} performs well and, with an AUC of 0.597, significantly better than a random predictor and the modifications discussed. CN_{IG} and WCN_{WIG}, perform slightly worse. The combined graph version WCN_{CG} performs comparably

Table 5.2 avg. AUC for CN

	AUC	SD	Sign.
CN_{SG}	0.597	0.082	0.001
CN_{IG}	0.554	0.061	0.005
WCN_{WIG}	0.560	0.097	0.037
WCN_{CG}	0.597	0.139	0.021

Table 5.3 avg. AUC for JC

	AUC	SD	Sign.
JC_{SG}	0.591	0.080	0.001
JC_{IG}	0.560	0.064	0.004
WJC_{WIG}	0.561	0.097	0.035
WJC_{CG}	0.582	0.125	0.027

Table 5.4 avg. AUC for AA

	AUC	SD	Sign.
AA_{SG}	0.597	0.086	0.001
AA_{IG}	0.563	0.070	0.005
WAA_{WIG}	0.571	0.094	0.014
WAA_{CG}	0.599	0.140	0.019

to CN_{SG}. Accordingly, we find that predicting edges in the *SG* by focusing exclusively on the *IG* or *WIG* does not increase the performance of *CN*. Moreover, predictions based on the *WIG* seem to outperform those based on the unweighted *IG*. This pattern is consistent across all $k \in \{100, 50, 20\}$, but only CN_{SG} is significantly different from 0.5 for all values of k.

We report the results for Jaccard's Coefficient in Table 5.3. Similar to *CN*, the traditional version of this metric (JC_{SG}) and its modification on the combined graph (WJC_{CG}) perform significantly better than a random predictor. We find further that JC_{IG} and WJC_{WIG} show a slightly worse performance, but still better than a random predictor. Only JC_{SG} shows an AUC value significantly different from 0.5 for all values of k.

Adamic/Adar shows AUC values close to 0.6 for its social graph version AA_{SG} and the combined graph version WAA_{CG} (see Table 5.4). Furthermore, AA_{SG} produces significant results for the top 100, top 50, and top 20 predictions. WAA_{CG} becomes non-significant for the top 20 approach. As already observed for Common Neighborhood and Jaccard's Coefficient, the *IG* and *WIG* versions of Adamic/Adar exhibit lower AUC values than the corresponding *SG* and *CG* variations and become non-significant for top 50 and top 20 predictions. Interestingly, for the top 50 and top 20 predictions, AA_{IG} and WAA_{IG} perform almost identically, suggesting a rather low performance for predictions based exclusively on properties of the *IG* structure.

Resource Allocation shows AUC values greater than 0.6 for RA_{SG} and WRA_{CG} (see Table 5.5) – the highest values observed in our experiment. While both performances are not significantly different from each other, they produce significant results for all top k predictions. Like AA_{IG} and WAA_{WIG}, RA_{IG} and WRA_{WIG} fail to produce significant results for $k \in \{50.20\}$.

Table 5.5 avg. AUC for RA

	AUC	SD	Sign.
RA_{SG}	0.606	0.091	0.001
RA_{IG}	0.553	0.053	0.002
WRA_{WIG}	0.570	0.074	0.003
WRA_{CG}	0.610	0.116	0.003

Table 5.6 avg. AUC for PA

	AUC	SD	Sign.
PA_{SG}	0.498	0.034	0.859
PA_{IG}	0.504	0.051	0.782
WPA_{WIG}	0.522	0.061	0.197
WPA_{CG}	0.509	0.042	0.450

Table 5.6 shows the results for Preferential Attachment. They suggest that *PA* does not perform significantly differently from a random predictor for all versions considered, showing AUC values very close to 0.5 for all values of k. In contrast to all other metrics, *PA* does not seem to lose but rather seems to gain a small degree of performance when moving from the *SG* to the *IG*. Considering weighted interaction seems to improve performance even more, while the combination of weighted interaction and the social graph information in the combined graph seems to decrease performance. Accordingly, we find that *PA* is the only metric of our set to perform better on the *IG* and *WIG* than on the *SG* and *CG*.

5.5   Discussion

The results presented in the previous section indicate that neighbor-based link prediction metrics perform well. Four of the five baseline metrics – Common Neighborhood, Jaccard's Coefficient, Adamic/Adar, and Resource Allocation – show a stable prediction quality above random for all $k \in \{100, 50, 20\}$. Only Preferential Attachment fails to produce valuable predictions. While all the neighbor-based metrics score rather close to each other, Resource Allocation and Adamic/Adar show the best results. Common Neighborhood and Jaccard's Coefficient follow closely. Even though the relative difference in the metrics' performances cannot withstand statistical tests, it remains stable for all top k predictions; a larger sample size would be required to improve the statistical accuracy of a relative comparison of the metrics. Our findings are consistent with those of Liben-Nowell and Kleinberg [15] and Lü and Zhou [17].

Compared to the baseline metrics, their applications to the *IG* prove to cause a loss of prediction quality. Four of five metrics lose a considerable amount of performance due to the loss of the *SG*'s information. Again, Preferential Attachment works differently. In contrast to the other metrics, *PA* gains performance on the *IG* – without producing significant predictions. Most of the metrics lose their significance for top 50 and top 20 predictions. Accordingly, considering only mutual interaction partners does not seem to be a promising approach to improve traditional link prediction.

Applying the weighted counterparts of the five baseline metrics to the weighted interaction graph has a similar effect. Except for Preferential Attachment, all metrics perform worse than their original versions. However, using the *WIG* for link prediction seems to work better than using the *IG* for all metrics considered, with the exception of *PA*.

The *CG* seems to be a more solid data source for the weighted neighbor-based metrics. All of them perform comparably to that of the baseline metrics. *RA* and *AA* seem to work especially well on the *CG*. Even though *RA* and *AA* perform minimally better on the *CG* than on the *SG*, the *SG* versions exhibit a better stability for smaller values of k. Considering the small differences in the performance of the *CG*- and the *SG*-based metrics, we find that using the *CG* does not improve the prediction quality to a significant degree.

As described by Wilson et al. [28], the *IG* represents rather strong connections between actors, whereas the social graph includes both strong and weak connections. Accordingly, when using *IG*-based metrics, we implicitly favor in our prediction pairs of actors with a strong indirect connection. The comparably low performance of *IG* and *WIG* metrics suggests that users of SNS befriend a considerable number of users, with whom they share a weak indirect connection. This also explains the good performance of *CG*-based metrics. Combining the *WIG* with the *SG* seems to compensate for the overrating of strong indirect connections, allowing the metrics to account for weak ones as well. Lü and Zhou [16] arrive at similar results. As already mentioned in Sect. 5.2, they applied weighted metrics to a variety of networks and concluded that the consideration of weights causes a bias towards strong tie predictions, and hence reduces the prediction quality by neglecting weak ties. In fact, Donath and Boyd [7] find that the cost of establishing new friendships on SNS are low. Hence, users tend to establish new friendships without considering the quality of the indirect connections to their targets. Our observations suggest this is true for the network considered. When it comes to establishing friendships in social networks based in social network sites, actors seem to favor quantity, not quality. This causes the simplest metrics to outperform more complex ones.

Despite these findings, interaction-based information has its value. We find that a considerable amount of actors in the analyzed network can be considered inactive and that activity is highly concentrated. We also find that the amount of recent interaction is correlated with the amount of link formation in the near future.

Beyond the scope of this work, it would be interesting to take a closer look at the construction of the combined graph. It provides a good starting point to adjust the degree to which interaction-related information is utilized in the prediction. In addition, a replication of this study with a larger sample size and comparable networks would be desirable to increase the overall validity. Moreover, one could analyze the number of identical predictions made by the metrics discussed to determine whether the weighted metrics are capable of capturing otherwise neglected friendship establishments.

5.6 Limitations

Before we conclude, we want to provide a brief summary of the most important limitations of this work.

First, we analyzed only one network. To increase the validity of our results, it would be necessary to replicate them on similar networks of large social network sites.

Second, our observation period was limited to 60 days. A longer observation period would enable a more detailed analysis of actors' interaction histories. Based on those histories, it would be possible to consider the dynamics and evolutionary aspects of user activity. Moreover, a longer test period would cover a larger number of link formations. In addition, observing user interaction for more than 60 days would decrease the degree of seasonal effects on our observations.

Third, it was necessary to limit the analysis to private communication. While we find this to be the primary form of interaction in the network studied, there are other forms of user interaction that could be of value to our approach. Accordingly, our results and findings are limited to a subset of activity. Considering other forms of activity is mandatory for further validation.

Finally and most important, the sample used in this work is small and has to be extended to ensure the validity of our results.

5.7 Conclusion

To improve traditional neighbor-based link prediction approaches, we questioned the value of using weighted metrics on variations of the social and interaction graphs of social network sites.

In cooperation with a German SNS operator, we performed an empirical experiment to compare the performance of traditional link prediction metrics with their weighted counterparts on different graphs generated by the SNS.

As a result, we find that traditional metrics applied to the social graph show the highest prediction quality. Our findings suggest that SNS users prefer quantity over quality in terms of establishing new connections. This results in the simplest metrics to perform best. They outclass more complex metrics that rely exclusively on interaction-related information. Our findings are consistent with existing studies. Furthermore, we find that despite the low performance of our metrics, a user's recent activity has a significant effect on his/her establishment of friendships in the near future. This is an important insight that highlights the relevance of previous interaction in the process of link formation.

References

1. Adamic L (2003) Friends and neighbors on the Web. Soc Networks 25(3):211–230
2. Backstrom L, Huttenlocher D, Kleinberg J, Lan X (2006) Group formation in large social networks: membership, growth, and evolution. In: Proceedings of the 12th ACM SIGKDD international conference on Knowledge discovery and data mining, ACM, New York, pp 44–54
3. Backstrom L, Leskovec J (2011) Supervised random walks: predicting and recommending links in social networks. In: Proceedings of the fourth ACM international conference on Web search and data mining, ACM, New York, pp 635–644
4. Barabasi A-L, Albert R (1999) Emergence of scaling in random networks. Science 286(5439):11
5. Börner K, Sanyal S, Vespignani A (2008) Network science. Annu Rev Inform Sci Technol 41(1):537–607
6. Boyd DM, Ellison NB (2007) Social network sites: definition, history, and scholarship. J Comput-Mediat Commun 13(1):210–230
7. Donath J, Boyd D (2004) Public displays of connection. BT Technol J 22(4):71–82
8. Fawcett T (2006) An introduction to ROC analysis. Pattern Recogn Lett 27(8):861–874
9. Gilbert E, Karahalios K (2009) Predicting tie strength with social media. In: Proceedings of the 27th international conference on human factors in computing systems – CHI '09, ACM, New York, p 211
10. Granovetter M (1983) The strength of weak ties: a network theory revisited. Social Theory 1:201
11. Hill R, Dunbar RIM (2003) Social network size in humans. Hum Nat 14(1):53–72
12. Kossinets G, Watts DJ (2006) Empirical analysis of an evolving social network. Science (New York, NY) 311(5757):88–90
13. Kumar R, Novak J, Tomkins A (2006) Structure and evolution of online social networks. In: Proceedings of the 12th ACM SIGKDD international conference on knowledge discovery and data mining – KDD '06, ACM, New York, p 611
14. Leskovec J, Backstrom L, Kumar R, Tomkins A (2008) Microscopic evolution of social networks. In: Proceeding of the 14th ACM SIGKDD international conference on knowledge discovery and data mining – KDD '08, ACM Press, New York, p 462
15. Liben-Nowell D, Kleinberg J (2007) The link-prediction problem for social networks. J Am Soc Inf Sci Technol 58(7):1019–1031
16. Lü L, Zhou T (2010) Link-prediction in weighted networks: the role of weak ties. EPL (Europhys Lett) 89(1):18001
17. Lü L, Zhou T (2011) Link-prediction in complex networks: a survey. Physica A Stat Mech Appl 390(6):1150–1170
18. Murata T, Moriyasu S (2008) Link-prediction based on Structural Properties of Online Social Networks. N Gener Comput 26(3):245–257
19. Newman MEJ (2001) Clustering and preferential attachment in growing networks. Phys Rev E 64(2):1–4
20. Newman MEJ (2001) The structure of scientific collaboration networks. Proc Natl Acad Sci U S A 98(2):404–409
21. Newman MEJ (2003) Mixing patterns in networks. Phys Rev E 67(2):1–13
22. Newman MEJ, Park J (2003) Why social networks are different from other types of networks. Phys Rev E 68(3):1–8
23. Panzarasa P, Opsahl T (2009) Patterns and dynamics of users behavior and interaction: network analysis of an online community. J Am Soc Inf Sci Technol 60(5):911–932
24. Roth M, Ben-David A, Deutscher D (2010) Suggesting friends using the implicit social graph. In: Proceedings of the KDD '10, ACM, New York, pp 233–241

25. Tylenda T, Angelova R, Bedathur S (2009) Towards time-aware link-prediction in evolving social networks. In: Proceedings of the 3rd workshop on social network mining and analysis – SNA-KDD '09, ACM Press, New York, pp 1–10
26. Watts DJ (2004) The "new" science of networks. Annu Rev Sociol 30(1):243–270
27. Watts DJ (2007) A twenty-first century science. Nature 445(7127):489
28. Wilson C, Boe B, Sala A, Puttaswamy KP, Zhao BY (2009) User interactions in social networks and their implications. In: Proceedings of the fourth ACM European conference on computer systems – EuroSys '09, Nuremberg, p 205
29. Xiang EW (2008) A survey on link prediction models for social network data. Ph.D. Dissertation, Department of Computer Science and Engineering, The Hong Kong University of Science and Technology
30. Zhou T, Lü L, Zhang Y-C (2009) Predicting missing links via local information. Euro Phys J B 71(4):623–630

Chapter 6
Integrated Modeling and Evolution of Social Software

Arnd Poetzsch-Heffter, Barbara Paech, and Mathias Weber

Abstract Social networks, public and private media, administrative and business processes are based on a multitude of interconnected information systems. These growing software infrastructures, called *social software* in the following, get more and more entangled with everyday life and the processes of the society. Social software and society together form complex sociotechnical systems. In this paper, we describe how such sociotechnical systems can be modeled in a way that integrates social and software processes. The models can be reflective in the sense that they include the processes for their own modification. This way, system evolution can be expressed and studied as part of the model.

6.1 Introduction

The growth of the World Wide Web is one prominent indicator for an increasingly strong interconnection of the real and virtual world. In this paper, we look at specific forms of such interconnections and argue that they have consequences for engineering future software infrastructures and, in particular, that extended modeling and evolution techniques are needed. We model sociotechnical software systems, STSS for short, as discrete systems that integrate the real-world stakeholders, processes and artifacts with the software processes and software artifacts. The models cover both runtime behavior and evolution steps. They can be reflective in the sense that they include the processes for their own modification. The goal is to provide a basis for a precise description of STSSs and for their evolution.

Since the beginning of programming, software has been used for many, quite different purposes. In particular, we distinguish between system software controlling the computer hardware and application software addressing the needs of users. Application software is often further classified into single-user applications,

A. Poetzsch-Heffter (✉) • M. Weber
University of Kaiserslautern, Kaiserslautern, Germany

B. Paech
University of Heidelberg, Heidelberg, Germany

© Springer International Publishing Switzerland 2014
K. Zweig et al. (eds.), *Socioinformatics - The Social Impact of Interactions between Humans and IT*, Springer Proceedings in Complexity,
DOI 10.1007/978-3-319-09378-9_6

embedded systems, and information systems. Traditionally, the focus of software engineering was on the analysis and design of the behavior of the software systems. With the growth of the information systems and their interdependence with the environment, techniques to understand and model user behavior and system environments became more and more important and an integral part of requirements engineering [4].

In the last 20 years, information systems together with the global communication facilities have become a crucial part of the infrastructure of the modern world. As *information infrastructure*, they link the different data sources and sinks together, stretch across many organizations, allow to realize systems of systems, have very large, heterogeneous user groups, and are increasingly intertwined with the processes of our societies. Accordingly, software engineering has developed methods for taking aspects into account that are not part of the running software system itself. For example, modern methods for stakeholder modeling go far beyond the classical *role separation* into *customer, developer,* and *user* and cover many aspects of who decides on the software and who is using it in which way [10]. Other aspects are the treatment of access control and ownership (see e.g. [14]). Even closer to our topic are so-called software eco-systems in which software product developers and third party developers collaboratively develop and evolve large software systems (see [7]). The engineering challenge will be to also include other social processes into the picture.

The relationship between social cooperation and software as an important infrastructure is investigated e.g. in [20]. They write: "We seek to explore the development and design of infrastructure; our main argument will be that a social and theoretical understanding of infrastructure is key to the design of new media applications in our highly networked, information convergent society." In [16], Pipek and Wulf build on the concept of infrastructuring and develop an integrated perspective on the design and use of information technology for IT infrastructures within organizations. Monteiro et al. [13] argue for a more open approach of computer-supported cooperative work (CSCW): "Information infrastructures are characterized by openness to number and types of users ..., interconnections of numerous modules/systems..., dynamically evolving portfolios of (an ecosystem of) systems..."

Existing and future social cooperation and processes will be built on the infrastructure consisting of an increasing number of interconnected information systems. Existing examples are trans-organizational information systems, conference systems, social networks, and other platforms, e.g., for urban management, crowdsourcing, and open source software development. For the future, we expect that also many administrative, legal, and political processes will be supported by software. In summary, this leads to an infrastructure for such social activities, called *social software* in the following, that combines data management, collaboration (possibly according to well-defined processes), communication, rights and identity management, and further aspects (as shown below). That is, we take a fairly broad view on social software.[1] For two reasons, the management and evolution of the resulting STSSs will become more and more complex:

1. Technical: The software part of the SSTS consists of many heterogeneous information systems.
2. Sociological: Whereas nowadays the responsibility for software development and maintenance is mostly in the hand of one organization or company, it will become more common that the responsibility will be shared among several *parties*, i.e., organizations, companies, or other legal entities.

In this paper, we describe a new integrated model for the design and multi-party evolution of social software. The main focus is on

– The modeling of aspects belonging to the environment of the software
– The relationships between the modeling entities
– The modification and evolution steps

We first characterize specific features of social software (Sect. 6.2). Then, we present and illustrate our model (Sect. 6.3). In Sect. 6.4, we consider the model-based evolution of social software. Section 6.5 concludes.

6.2   Social Software and Its Characteristics

Social software supports processes among real world stakeholders. The processes might consist of well-defined collaboration steps or might just be some open cooperation. As for business processes [21], such *processes* can be described in terms of

– The *stakeholders* (individuals, organizations) and their relationships
– The basic activities and *tasks* performed by the stakeholders
– The *data* and *artifacts* managed and produced by the processes
– *Policies* and rules governing the ownership, access, and modification rights

The mentioned policies and rules might only be designed for the social software under consideration. For example, in Wikipedia, they are under the control of the organization running Wikipedia. However, in social software that is deeper linked into existing processes of the society, the rules might be decided by stakeholders outside of the software development and maintenance cycles. In particular, changes in legal *regulations* or laws can directly call for modifications of the software systems supporting the related administrative processes.

Characteristics. From a technical point of view, social software is essentially about distributed communication and information systems that support coordinated

---

[1]E.g. compared to [18].

behavior in form of partially predefined, adaptable processes ([21] explains four different degrees of framing). The interesting characteristics of social software are of a nontechnical nature:

- Stakeholder communities: For social software, it is often crucial that they trigger the formation of stakeholder communities (e.g., Wikipedia with only one author or a crowdsourcing platform with only one user do not work). This community building is a voluntary process in contrast to, e.g., enterprise resource planning systems where the use is enforced by organizations.
- Heterogeneous stakeholders: Social software can span different organizations and typically has many groups of stakeholders. The stakeholders might be involved in development, use, and evolution of the software; but there might also be stakeholders that do not know at all about the social software (e.g., committees deciding on relevant regulations).
- Infrastructure: Social software shares properties with other kinds of technical infrastructure. In particular, it is intended to be long-living and to evolve in mutual interdependence with the surrounding conditions in society.
- Compliance: To support social processes that are based on, extend or combine existing administrative, legal, or business processes, the rules and regulations of these processes are defined by committees or bodies outside the influence of the software development. Modification of the rules might break the compliance of the software with the regulations.

These characteristics are not meant to yield a sharp classification into social and other software. And we do not expect that for every social software all of the criteria are of crucial importance (e.g., compliance is not so problematic for Wikipedia).[2] But the aspects get more complex in the context of social software.

Thus, we argue here that these aspects should play a more prominent role in the engineering and evolution of social software and that extended modeling techniques are needed. The central argument is that these four characteristics result in new challenges at the interplay between society and software. Let us look at some of these challenges:

- If stakeholder communities are important, then the systems have to support *mechanisms for flexible adaptation* to stakeholders' needs. However, managing such adaptations is a challenge, because many of these systems have very large user groups and, more importantly, may have many stakeholders that would like to participate in decisions on the adaptation.
- A simple role separation into user, developer, and customer is no longer adequate or possible. In particular, users can modify data that affect the work of others (like e.g., in Wikipedia), can grant access rights to other user groups (like e.g., in controlled forms of open source software development), and might even participate in the evolution of the social software (cf. [9]). Thus, a more *detailed*

---

[2]Except maybe for copyright regulations.

and dynamic management of rights and policies to handle conflicts is needed. Whereas in classical software engineering resolution of conflicts among user groups is essentially part of the requirements engineering, social software needs policies and mechanisms to also resolve conflicts at run time (e.g., Wikipedia needs a mechanism to accept articles).

- If a software system can run into compliance problems, the *parties* (i.e., committees, bodies, organizations, etc.) deciding on the rules underlying the software should be considered as stakeholders and should be "warned" that modification of the rules might entail expensive modification of the software. More importantly, if a software system realizes processes based on regulations of several, so far independent parties a process for joined decision taking for these parties has to be set up as part of the software life cycle management.

We are interested in modeling of social software for which such challenges have to be met simultaneously. In particular, we assume that there is no central decision body and that the software is managed in a distributed way.

Examples. An example of multi-party social software is Hochschulstart, an ongoing project that is aiming to build a system for the application and enrollment of students at German universities. The system focuses on subjects where the universities only offer a limited number of places. The system allows students to apply at several universities. Each university independently decides about its applications according to its local rules. If a student is admitted by a university and accepts the offer, his/her possibly successful applications at other universities are canceled. Places that become available by such cancellations are offered to other applicants. Hochschulstart is a multi-party system that has suffered from the large number of universities and ministries involved in the requirements analysis and development (see [12]). It also illustrates the compliance problems as well as the resulting development challenges. Another prominent example of a multi-party system that has been much discussed in the German public is the electronic health card (Elektronische Gesundheitskarte).

In the following, we will use a fictitious campus management system, called FCMS, to illustrate aspects of our model. The system realizes the management and support of the courses and exams:

- FCMS is a multi-party system, because the university departments are fairly independent in deciding on the courses that are offered and on the study plans, whereas the university decides on the examination regulations which itself have to be consistent with the university law on the state level.
- FCMS should be flexible in handling many different aspects, e.g., enrollment into courses, examination administration, project management, student progress, etc. and be adaptable to new aspects (e.g. online exams).
- FCMS illustrates compliance as the software has to comply to the underlying rules and regulations.

FCMS also illustrates another important aspect, namely the continuous transition from completely manual, paper-based management (as in the old days) to sophisticated computer support that allows optimal management of lecture halls, minimization of course overlaps, and checking whether the cancellation of a course might endanger the study plans of a larger number of students.

6.3 An Integrated Model for Social Software

To meet the sketched engineering challenges, we need a better understanding of SSTSs. In particular, we have to emphasize the relationships between the aspects directly related to the software (the software itself and the data managed by the software) and those aspects in the social environment that are relevant for development, use, and evolution of the software (stakeholders, legal regulations, policies, etc.). To achieve this, we model SSTSs as reflective transition systems. The model integrates the processes that are supported by social software interacting with users with the processes that are performed manually by the involved stakeholders. Figure 6.1 outlines the modeling elements and their associations as a UML class diagram.[3] SSTSs are modeled as closed and modifiable systems, i.e., the model

- contains all relevant information for the dynamic behavior of the sociotechnical system (closed), and
- supports mechanisms for modifying all parts of the system.

The basic entities of such systems are *stakeholders*, *data*, *processes*, and *artifacts*. Artifacts are specific models, descriptions and software defining the possible states and allowed behaviors of the system as explained below. The meta-model essentially extends the system concept model of [15] by the artifacts in order to capture the evolution of the artifacts as part of the system evolution.

System states and semantics. Before we further motivate the different model entities and investigate their relationships, let us consider the semantical idea underlying the approach. A sociotechnical system is modeled as a (complex) *state transition system*. The states have four components:

- The current stakeholders, their roles, and relationships: A state change could be, e.g., that a new user is registered, that users change their role or enter a new group, that a software company stops working for the system, i.e., is no longer a stakeholder, etc. The stakeholders and their relationships must conform to the *stakeholder model*. If changes should be done that do not conform to the current stakeholder model, the stakeholder model must be modified first.
- The current data: The data can be complex and distributed; it can be managed manually or by computers. It must conform to the *data model*; data changes must maintain the integrity constraints of the data model.

---

[3]For better readability, we use different line formats for associations and emphasize the directions of the associations by arrowheads.

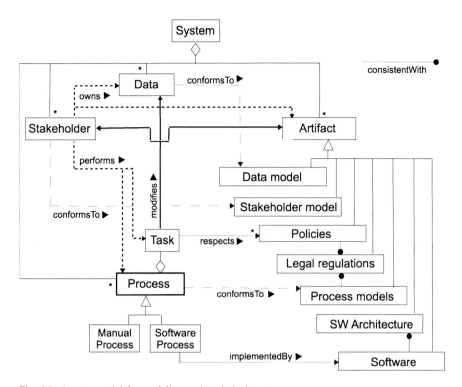

Fig. 6.1 A meta-model for modeling sociotechnical systems

- The current artifacts: as shown in Fig. 6.1, we assume at least seven artifacts: the stakeholder and data model, a description of the policies, a list of relevant legal regulations, the process models, the description of the software architecture, and the software itself. The artifacts belong to the state. State changes on the artifacts modify the structure and behavior of the STSS. In particular, software maintenance and evolution steps correspond to artifact modification (as discussed below).
- The currently executing processes with their current execution states: a process is manually performed or software-supported (see Fig. 6.1). It consists of one or several *tasks*. The tasks and processes are performed by a stakeholder and have to conform to a *process model*.

We assume that state changes can *only* happen by executing a task, i.e., a task can modify the data (normal behavior of an information system), the set and grouping of the stakeholders and the artifacts. Tasks can be executed concurrently, but we assume that they have the ACID properties, i.e., they are executed atomically in an isolated way, maintain consistency of the system model, and guarantee durability [8].

It is important to understand the sociotechnical nature of the system model. In general, the different entities can be part of the real world and/or be part of the virtual world managed by the social software. For example, certain stakeholders might not be managed by the software whereas others are an important aspect of the software functionality. In our FCMS, e.g., the faculty councils are considered to be stakeholders as they decide on legal regulations[4] underlying and influencing the FCMS; they are not treated in the FCMS software. On the other hand, enrolled students are both part of the real world and of the virtual world "realized" by the FCMS software. Similarly, a process can be performed by some stakeholders without any computer support (*manual process*) or by software.

Stakeholders. In our model, a stakeholder is an individual, group, or organization that is relevant for the regulations, development, use, and evolution of the sociotechnical system. Thus, similar to [2], we take a fairly broad view.[5] Stakeholders of our FCMS are, e.g.,

– Users: students, professors, research assistants, administrative personal, etc.
– System administrators, managing the social software
– Developers: partially in an external company, partially in the university
– Financier: typically the departments and the university
– Regulation authorities: deans, faculty council, university senate, etc. as they define the regulations underlying some of the FCMS's processes
– System specific committees, e.g. to manage the evolution of the FCMS

The stakeholder model does not depend on other entities of the system, but is used in the other models; e.g., stakeholders trigger and perform processes. They are the owners of data and artifacts where the ownership relation also expresses a form of responsibility. We do not favor a specific technique for stakeholder analysis and modeling. A possible approach is described in [11].

Processes and tasks. The system activities are performed by processes. A process consists of tasks where a task describes an atomic transition in the system. We assume that all processes have to conform to a process model/description.[6] That is, we require to describe even the manual processes, at least if they comprise tasks that modify the state of the system. Process descriptions can be very simple (consisting just of a single task) and abstract (just a name or informal sketch what they do). Of course, they can also be very detailed. The tasks are associated with the policies that they have to respect. A process and its tasks are performed by one or several stakeholders. The stakeholders must have sufficient rights to perform the tasks as defined by the policies.

Processes have very different impact. They can change user data in the system. They can add or remove stakeholders. They can modify the software architecture

---

[4] Such as the form of study plans.

[5] Different from [2], we relate the stakeholders to an STSS and not to a product.

[6] We use the terms "process model" and "process description" synonymously.

and the software (for maintenance or evolution). They can also modify all other artifacts. For example in the FCMS, there might be processes for changing the offered courses of a department or for modifying study plans. Typically, the stakeholders of such processes are legal entities, for example the faculty council.

Data and artifacts. The data captures the information that is needed to execute the manual and software processes. The data model describes the structure and integrity constraints for that data. In addition to ground data, the data model can cover meta-data or reify some aspects/parts of the other models. E.g., the data model might express part of the stakeholder model to manage users or it can be used to maintain the dynamically changing ownership relations.

In social software, it is important to clarify who has the right to access and modify which parts of the data and which artifacts. This is not only important during normal execution, but in a multi-party setting also for maintenance and evolution. Our integrated modeling approach addresses this requirement in the following way:

– The subjects, sometimes also called principals (cf. [17] and [1]), that can have rights are stakeholders and are thus modeled by the stakeholder model.
– The entities that can be accessed and modified are data elements, artifacts, and the groups and roles of stakeholders.
– As a basis to formulate policies, we model an ownership relation between stakeholders and the data elements and artifacts (see the "owns" association in Fig. 6.1). For the processes and tasks, the performs association also expresses ownership. Thus, we assume that every entity in the system except for the stakeholders has an owner. The owner can be considered to be the responsible stakeholder for the entity.
– The concrete access and modification rights are formulated as policies. The tasks being the only entities that perform accesses and modifications have to respects these policies.
– As policies are artifacts of the system, they are subject to change.

In practice, policies can be of a quite different nature. Some of them are closely linked with the stakeholder and data model. For example in FCMS, there might be policies defining who is allowed to enter the grades for a specific course. Other policies are of a more general nature. For example, the policies defining the rights to modify the stakeholder model or the legal regulations. There could also be policies that allow the transfer of rights from one stakeholder to others (cf. e.g. [22]). Policies built on stakeholder models and are important for security in social networks (see e.g. [6]) and in e-commerce software (see e.g. [5]).

Policies and processes have to be consistent with legal regulation (cf. Fig. 6.1) that are typically decided and managed by committees and boards that are outside the development of the social software. In our integrated model we consider the legal regulations as artifacts within the model and the committees and boards as stakeholders. This allows us

– to make the consistency requirements between the policies, processes and regulations explicit and

– to model the modifications of the regulations as processes that are part of the sociotechnical system.

The software architecture and the software itself are also artifacts of the model. The software implements some of the processes and has to conform to the process model (and the underlying regulations if any). Consequently, it also has to respect the policies. Software maintenance and evolution corresponds to artifact modification and is thus an integral part of the modeled sociotechnical system.

Integration and integrity. The models of STSS are *integrated* in a static and dynamic sense:

– Static integration means that the different entities and artifacts of the model refer to each other. For example, stakeholders perform tasks, tasks respect policies, and policies are expressed using roles of the stakeholder model.
– Dynamic integration means that the approach integrates normal execution, maintenance, and evolution (cf. Sect. 6.4).

The integration enables the precise formulation of dependencies and integrity constraints among the entities and artifacts. In particular, it allows *reflective* modeling. For example, a policy can grant the right to modify itself to one of the groups mentioned in the stakeholder model. Thus, the artifacts become first-class objects of the evolution process. This is similar to legislative systems that comprise laws about processes on how to modify the law.

6.4 Evolution of Social Software

Social software and the surrounding sociotechnical system are subject to change. We distinguish three kinds of changes and relate them to our modeling approach:

– *Normal execution* corresponds to the use of the social software. It only affects the data, the state of the processes and the stakeholders (data might be modified, processes make progress, users might be added or removed from the system). It does not touch the artifacts.
– *Software maintenance* corresponds to those changes that only affect the software and those parts of the artifacts that are related to the software. E.g., software maintenance can change the data model w.r.t. the data managed by the software. Software maintenance corrects errors, adapts to the technical platforms (e.g., a new operating system) and improves performance, usability or other nonfunctional attributes.
– *Evolution* refers to mutual changes of software and its social environment; for example, supporting additional processes by software, reacting to new or different functional requirements, possibly affecting all artifacts of the sociotechnical system.

Many authors do not distinguish between software maintenance and evolution (see e.g. [19], Sect. 4.3.4.) or use the terminology in a slightly different way (see, e.g., [3]). Evolution here means that the social environment and the software changes at the same time.

Evolution is complex, in particular if an agreement between different stakeholders has to be reached (multi-party). In our model, evolution can be modeled as an (internal) process of the sociotechnical system. To illustrate how evolution can be handled in our model, we consider two different scenarios:

- *Planned evolution*: We can use our model to incorporate anticipated evolution processes as part of the process model. For example in the FCMS, we might foresee future modifications of the examination regulations. We can define a (mainly manual) process how changes of the regulations are implemented by adapting the software. In particular, we can establish specific stakeholders (typically groups of employees) who drive this process and develop and deploy the software updates. Another example of planned evolution is a process that collects user feedback and requests for new features, organizes a committee to decide which requests should be realized, and triggers the implementation. Of course, such planned processes can be (partially) supported by the social software itself.
- *Unplanned evolution*: In our model, unplanned evolution means that the (current) process model does not have a process tailored to handle the desired evolution steps. Unplanned evolution can be supported in a more or less structured way:
  - Unstructured: The process model supports a process that essentially enables the modification of all entities, in particular all artifacts, in an arbitrary way. Of course, only specific stakeholders should be allowed executing this powerful process. This corresponds to superuser rights in the area of operating systems.
  - Structured: The process model supports a process that permits to extend the process model by a new process for the (unplanned) evolution steps. A structured approach is in particular appropriate if many stakeholders are involved in the evolution.

In summary, our model supports evolution of STSS in ways that meet the specific needs of the evolution scenarios. The model of the sociotechnical system provides a framework for designing evolution steps and for better mastering the complexity of evolutions.

6.5   Conclusion

We described an approach to model sociotechnical systems as closed, discrete, and reflective systems that integrate the software and its social environment. Like with self-modifying programs, the models comprise the artifacts that describe the system as part of the system itself. By modifying these artifacts, the system can change its structure and behavior. The self-modifications allow expressing the planned and unplanned system evolution in a controlled and flexible way. To achieve this, it is not enough to have rules describing how social software works and how it can be evolved. We also need rules about the evolution of the rules.

In the current stage, the approach is meant as a step towards a better understanding of STSSs and as a mental background for software engineers that aim to analyze development and evolution steps of social software.

References

1. Abadi M (2003) Logic in access control. In: LICS, Ottawa. IEEE Computer Society, Los Alamitos, p 228
2. Alexander IF, Robertson S (2004) Understanding project sociology by modeling stakeholders. IEEE Softw 21(1):23–27
3. Bennett KH, Rajlich V (2000) Software maintenance and evolution: a roadmap. In: Finkelstein A (ed) ICSE – future of SE track, Limerick. ACM, pp 73–87
4. Cheng BHC, Atlee JM (2007) Research directions in requirements engineering. In: Briand LC, Wolf AL (eds) FOSE, Minneapolis, pp 285–303
5. Cheng EC (2000) An object-oriented organizational model to support dynamic role-based access control in electronic commerce. Decis Support Syst 29(4):357–369
6. Chronopoulos K, Gouseti M, Kiayias A (2013) Resource access control in the facebook model. In: Abdalla M, Nita-Rotaru C, Dahab R (eds) CANS, Paraty. Volume 8257 of Lecture notes in computer science. Springer, pp 179–198
7. Dittrich Y (2014) Software engineering beyond the project – sustaining software eco-systems. Inf Softw Technol 56(11):1436–1456
8. Härder T, Reuter A (1983) Principles of transaction-oriented database recovery. ACM Comput Surv 15(4):287–317
9. Ko AJ, Abraham R, Beckwith L, Blackwell AF, Burnett MM, Erwig M, Scaffidi C, Lawrance J, Lieberman H, Myers BA, Rosson MB, Rothermel G, Shaw M, Wiedenbeck S (2011) The state of the art in end-user software engineering. ACM Comput Surv 43(3):21
10. Lim SL, Finkelstein A (2012) StakeRare: using social networks and collaborative filtering for large-scale requirements elicitation. IEEE Trans Softw Eng 38(3):707–735
11. Lim SL, Ncube C (2013) Social networks and crowdsourcing for stakeholder analysis in system of systems projects. In: SoSE, Maui. IEEE, pp 13–18
12. Mertens P (2012) Schwierigkeiten mit IT-Projekten der Öffentlichen Verwaltung – Neuere Entwicklungen. Inform Spektrum 35(6):433–446
13. Monteiro E, Pollock N, Hanseth O, Williams R (2013) From artefacts to infrastructures. Comput Support Coop Work 22(4–6):575–607
14. Ni Q, Bertino E, Lobo J, Brodie C, Karat C-M, Karat J, Trombetta A (2010) Privacy-aware role-based access control. ACM Trans Inf Syst Secur 13(3):24:1–24:31
15. Paech B (2000) Aufgabenorientierte Softwareentwicklung – integrierte Gestaltung von Unternehmen, Arbeit und software. Springer, Berlin/New York
16. Pipek V, Wulf V (2009) Infrastructuring: toward an integrated perspective on the design and use of information technology. J AIS 10(5):306–332
17. Sandhu RS, Coyne EJ, Feinstein HL, Youman CE (1996) Role-based access control models. IEEE Comput 29(2):38–47
18. Pacuit E, Parikh R (2006) Social interaction, knowledge, and social software. In: Goldin D, Smolka SA, Wegner P (eds) Interactive Computation. Springer, Berlin/Heidelberg, pp 441–461
19. Sommerville I (2007) Software engineering, 8th edn. Addison-Wesley, London/Reading

20. Star SL, Bowker GC (2002) How to infrastructure. In: Lieverouw LA, Livingstone S (eds) Handbook of new media: social shaping and social consequences of ICTs. SAGE, London/Thousand Oaks, pp 151–162
21. van der Aalst WMP (2013) Business process management: a comprehensive survey. ISRN Softw Eng 2013:37p
22. Wainer J, Kumar A, Barthelmess P (2007) DW-RBAC: a formal security model of delegation and revocation in workflow systems. Inf Syst 32(3):365–384

Chapter 7
Social Network Analysis in the Enterprise: Challenges and Opportunities

**Valentin Burger, David Hock, Ingo Scholtes, Tobias Hoßfeld,
David Garcia, and Michael Seufert**

Abstract Enterprise social software tools are increasingly being used to support
the communication and collaboration between employees, as well as to facilitate
the collaborative organisation of information and knowledge within companies. Not
only do these tools help to develop and maintain an efficient social organisation, they
also produce massive amounts of fine-grained data on collaborations, communication and other forms of social relationships within an enterprise. In this chapter, we
argue that the availability of these data provides unique opportunities to monitor
and analyse social structures and their impact on the success and performance
of individuals, teams, communities and organisations. We further review methods
from the planning, design and optimisation of telecommunication networks and
discuss challenges arising when wanting to apply them to optimise the structure
of enterprise social networks.

7.1 Introduction

We are currently witnessing a rapidly increasing adoption of technical systems in
numerous aspects of everyday life. In particular, the widespread use of information
and communication technologies in which interactions and collaborations between
humans are an integral part has led to the rise of so-called *socio-technical* systems.
A defining characteristic of these systems is that they consists of interwoven social
and technical layers, which both are crucial for their functioning. Many examples
for such socio-technical systems, like, e.g., social media platforms, collaborative
web applications have recently gained popularity. The popularity and success of

V. Burger (✉) • D. Hock, T. Hoßfeld, M. Seufert
Institute of Computer Science, University of Würzburg, Würzburg, Germany
e-mail: valentin.burger@informatik.uni-wuerzburg.de; david.hock@informatik.uni-wuerzburg.de;
hossfeld@informatik.uni-wuerzburg.de; seufert@informatik.uni-wuerzburg.de

I. Scholtes • D. Garcia
Chair of Systems Design, ETH Zürich, CH-8092 Zürich, Switzerland
e-mail: ischoltes@ethz.ch; dgarcia@ethz.ch

© Springer International Publishing Switzerland 2014
K. Zweig et al. (eds.), *Socioinformatics - The Social Impact of Interactions between
Humans and IT*, Springer Proceedings in Complexity,
DOI 10.1007/978-3-319-09378-9_7

these platforms has resulted in the adoption of similar technologies in an enterprise context. Specifically, *enterprise social software tools* are increasingly being used to support the collaboration between employees, as well as to facilitate the collaborative organisation of information and knowledge. Notable examples include groupware systems, collaborative information spaces like Wikis or Blogs, instant messaging, project and knowledge management platforms and – increasingly – social networking services specialised for an enterprise context. While these systems serve different purposes, they have in common that they generate massive amounts of fine-grained data on collaborations, communication and other forms of social relationships between employees. On the one hand, the availability of such data introduces severe privacy issues and thus raises a number of ethical challenges that urgently need to be addressed. On the other hand, such data provide interesting opportunities to gain insights into the structure and dynamics of the social organisation of an enterprise. Not only can important individuals be identified that otherwise may go unnoticed, a monitoring of evolving social structures by means of quantitative measures may also help to identify problems and take adequate countermeasures. A study of quantitative performance indicators – which are often available in an enterprise context – can furthermore provide unique insights into the effect of social structures on the success and performance of individuals, groups and projects.

In this chapter, we argue that the monitoring and optimisation of enterprise social networks provide interesting perspectives for a social informatics research agenda. Intuitively, one could argue that an optimisation of social networks is not easily possible, since they emerge in a self-organised way and thus cannot be influenced or designed. While this is true in many social systems, knowledge from the *planning, design and optimisation of telecommunication networks* can nevertheless be used to analyse the efficiency and resilience of social networks. Furthermore we argue that – through a targeted structuring of teams, the introduction and configuration of communication and collaboration tools as well as the design of corporate policies – the evolution of social networks in an enterprise can – at least to a certain extent – be influenced and shaped. Knowledge from network design may thus be utilised in emerging social organisations to improve resilience and to optimise their efficiency. Similarly, decisions in the design of socio-technical systems can influence the structure of social organisations into which they are embedded.

This chapter is structured as follows: In Sect. 7.2 we provide an overview of measures used in the analysis of complex networks, and interpret their meaning in the context of enterprise social networks. In Sect. 7.3, we show how one can use these measures to monitor the structure and evolution of social networks extracted from socio-technical systems. Section 7.4 introduces basic notions used in the optimisation of communication networks and discusses how these approaches can be used in the context of social networks. We further highlight research challenges arising in the modelling, analysis and optimisation of enterprise social networks. Highlighting links between research in the fields of *network planning and design* and *social informatics* – which are currently not well integrated – we conclude in Sect. 7.5.

7.2 Quantitative Analysis of Complex Networks

The increasing availability of data that allows to reconstruct networks of interactions between elements in complex systems has led to a massive surge of interest in the quantitative analysis of complex networks. During the last few decades, a comprehensive set of measures has been introduced, which allow to quantify characteristics of complex networks. Referring to available reference books for more details [10, 17], in the following, we provide a brief overview of these measures and interpret their meaning in the context of enterprise social networks. We particularly categorise measures into *node-centric* measures, which are targeted at capturing characteristics of individual nodes, as well as *network-centric* measures, which capture systemic properties of complex networks. In the following, we refer to a network $G = (V, E)$, which consists of a set V of *nodes* as well as a set $E \subseteq V \times V$ of links that interconnect nodes. In the context of enterprise social networks, we commonly assume that nodes represent *employees* or *co-workers* within an enterprise, while links between them are thought to represent some form of *social interaction*, like, e.g., an exchange of information, a conversation across E-Mail, instant messaging or voice communication services, or the collaboration in the context of a particular project. The ability to automatically reconstruct enterprise social networks require data on these interactions to be recorded, which typically implies that they are mediated via some type of technical system. However, the increasing adoption of modern wearable computing and sensing technologies highlights scenarios where networks can also be constructed from direct interpersonal communication between employees as well as their mobility traces [11].

For the remainder of this section, unless stated otherwise, we assume that networks are *undirected*, i.e. a link (v, w) between two nodes v and w implies that the reverse link (w, v) exists, in which case both links can be conveniently represented by a single undirected link. Even though the number, frequency or intensity of recorded social interactions can often be used to establish a notion of *link weights*, for the sake of simplicity, we further assume that networks are unweighted, i.e. the weight or strength of all links is the same. A simple example for such an undirected, unweighted network – representing social interactions between members of a software development team – is shown in Fig. 7.1.

7.2.1 Node-Centric Metrics: Centrality and Topological Embedding

A basic task in the analysis of complex networks is to quantify the importance – or centrality – of individual nodes, as well as how they are embedded in the overall topology. In the following, we thus give a brief overview of different measures that have been proposed for this purpose, and how they can be interpreted in the context

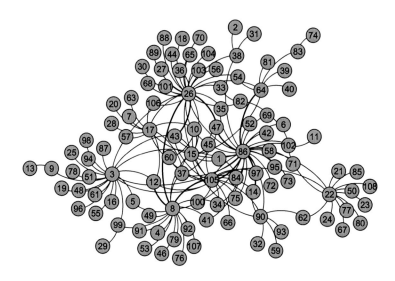

Fig. 7.1 An example network $G = (V, E)$ consisting of a set of *nodes* V and *links* $E \subseteq V \times V$

of enterprise social networks. For their interpretation, it is important to consider the semantics of links and the resulting meaning of the network topology in the given context.

Node degree. A particularly simple measure which is often used to capture the importance of a node is its *degree*, which is defined as the number of direct neighbours to which it is connected. A natural tendency of (social) networks occurring in many contexts is that they exhibit *heavy tailed distributions* of node degrees, implying that there are a few nodes whose degrees are magnitudes larger than the degrees of the majority of nodes in the network. In the context of enterprise social networks where links represent collaborations, an exchange of information or communication, the degree of nodes can be used as the most basic proxy for the popularity or importance of the persons they represent. While heavy-tail degree distributions arise naturally in social networks, they can be used to evaluate the *centralisation* of the social organisation of collaborating teams. Furthermore it has been argued that the cognitive capabilities of humans pose a limit to the number of stable inter-personal social relations [2]. Thus, structures in which most of the links are concentrated on only a few nodes, can indicate situations in which central employees are being overburdened by communication, This can possibly have negative consequences for the efficiency of a social organisation. Furthermore, random networks with heavy-tail degree distributions have a tendency to be vulnerable against the loss of high-degree nodes, meaning that the network can be disconnected even though only a small fraction of its most connected nodes are removed [1]. As such, the degree of centralisation of an enterprise social network in terms of node degrees can be seen as a simple proxy for the resilience of a social organisation against the loss of its most connected members.

Path-based centrality measures. A different set of measures for the importance of nodes in a network are those which are based on the topology of *shortest paths* between nodes in a network. One important example is the *betweenness centrality* of a node v, which is defined as the number (or fraction) of shortest paths between any pair of nodes that pass through node v [4]. Similarly, the *closeness centrality* of a node v is defined based on the average distance of a node v from any other node in the network. To obtain a measure of centrality in which higher values indicate more central nodes, the inverse of the average distance is typically used, meaning that a node with closeness centrality 1 is directly connected to any other node, while its closeness centrality tends to zero, as the average distance to other nodes tends to infinity. Instead of the average distance to all other nodes, one can alternatively study the maximum distance of a node v to any other node in the network, which is called its *eccentricity*. The betweenness and closeness centrality of nodes in an example network, as well as their eccentricity is depicted in Fig. 7.2a–c.

While nodes with high degree have a tendency to be important also in terms of path-based centrality measures, this correlation does not hold necessarily. Nodes with high degree can still be in the periphery of a network in terms of their average or maximum distance to all other nodes, meaning that they have small eccentricity and closeness centrality. Conversely, nodes with small degree can nevertheless reside at the core of a network through which many paths pass, meaning that they have high betweenness centrality. Path-based centrality measures thus capture a different dimension of topological importance and can thus play an important role in the monitoring, analysis and optimisation of enterprise social networks. In particular, individuals with high betweenness centrality may go rather unnoticed as they are not necessarily in contact with *many* colleagues. Nevertheless, their loss will have considerable impact on the flow of information, as it will change a seizable fraction of shortest paths between other nodes in the network. Furthermore, individuals with high betweenness centrality often play the role of *mediators*, which interconnect different parts of an organisation and bridge information between different communities. At the same time, a highly skewed distribution of betweenness centrality can be interpreted as a sign of high centralisation, which potentially poses a risk for efficiency and resilience of social organisations. While betweenness centrality captures shortest paths *passing through* a node, closeness centrality and eccentricity focus on the length of paths *starting or ending in a node*. As such, they capture how individuals are able to receive and propagate information travelling across shortest paths: Individuals with high closeness centrality can be seen as good information spreaders, since they can propagate information throughout the network most quickly. Nodes with small closeness centrality on the other hand are on in the periphery of a social organisation, thus receiving information – on average – later than others. Similarly, for nodes with high eccentricity there exist other nodes in the network that can only be reached via long paths. Individuals that play a central role

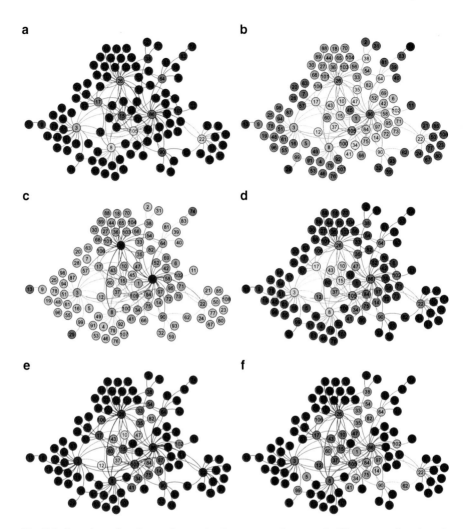

Fig. 7.2 Overview of node-centric metrics in an example network. The score of each node is depicted by its colour (*red*: low, *yellow*: medium, *blue*: high). (**a**) Betweeness centrality. (**b**) Closeness centrality. (**c**) Eccentricity. (**d**) Eigenvector centrality. (**e**) Clustering coefficient. (**f**) Coreness

in a social organisations should thus – in general – exhibit high closeness centrality and small eccentricity.

Clustering coefficient. Apart from different dimensions of importance introduced above, another important characteristic of nodes is how they are embedded into the topology of a network. One interesting aspect is, for instance, whether the neighbours of node v are also connected to each other, or – in other words – whether *triads* $(v, x), (v, y)$ around a node v are closed. To quantify this property, the *clustering coefficient* of a node v is defined as the fraction of pairs of neighbours x and y of v for which a link (x, y) exists. The clustering coefficient of nodes in a sample communication network is visualised in Fig. 7.2e. In the context of enterprise social networks, several different interpretations for the presence of nodes with high clustering coefficient are possible: First of all, naturally evolving social networks are known to have a – compared to random networks – high average clustering coefficient. At the same time, they exhibit a small diameter that is due to so-called *weak ties* which bridge the local cluster structures around nodes. Social networks with such a combination of high clustering coefficient and small diameter are usually called *small worlds*. Different from general networks with low diameter, small worlds typically have the property that they are *navigable* for humans, i.e. individuals are able to locally route information along short paths without global knowledge about the network topology. One property that enables individuals to quickly identify neighbours which lie on short paths to a given target is *funneling*, i.e. the fact most short paths pass only through a small set of neighbours which have connections outside local cluster structures. As such, the clustering coefficient of enterprise social networks can be used to quantify aspects that influence their navigability, an important property for the routing of information. Being aware which colleagues represent weak ties to other communities (and which thus transcend local clustering structures) is likely to be important, e.g. in order to quickly identify which colleagues have a particular expertise or work on similar projects, even if they are not directly connected to an individual. Furthermore, a high clustering coefficient of a node can be used as a proxy for the impact of removing this individual from a team: For a node v with high clustering coefficient, most of v's neighbours can still communicate or collaborate with each other even if v is removed from the network. Similarly, a high clustering coefficient can help to mitigate the overload of a central node v, since communication between two neighbours u and w can alternatively bypass v via a direct link (u, w).

Coreness. Another aspect of the embedding of individual nodes into the topology of a network is captured by their *coreness*. The k-core of a network is defined as the largest subgraph of a network, in which each node has a degree of at least k. Based on this decomposition in different k-cores, the *coreness* of a node is the maximum k-core to which it belongs. The coreness of nodes in the example network is shown in Fig. 7.2f. In particular, the k-core of a network is the largest connected component which is left when repeatedly removing all nodes with degree smaller than k. As such, the k-core decomposition of a network, as well as the coreness of nodes, plays an important role in the analysis of resilience of social network structures against cascading processes. The presence of k-cores with high k in an enterprise social network can be related to its ability to withstand *turnover* of employees, as well as potential cascades or network effects potentially triggered by individuals leaving a company.

7.2.2 Network-Centric Metrics: Resilience and Efficiency

Apart from measures that address the importance and topological embedding of individual nodes, an important further contribution of network theory is the provision of *aggregate, network-centric measures* that can be used to capture *systemic properties* of complex networks. In the following, we briefly introduce a set of network-centric measures that can be related to two particular systemic properties of complex networks: their resilience against failing nodes or links as well as their efficiency in terms of information propagation. We then interpret them in the context of enterprise social networks.

Network size, compactness and average degree. The simplest possible aggregate quantities of a complex network are the number of its nodes and edges. Based on these quantities, the *compactness* of a network can be defined as the ratio between the number of edges and the number of edges that could possibly exist in a network with the same number of nodes. The *average degree* is defined as the average of the degrees of all nodes. For networks with scale-free, heavy-tail degree distributions, the average degree is – in general – not a good representation for the typical degree of connectivity in the system. In such networks the degree of most nodes is in fact much smaller than the average, while a few nodes have degrees orders of magnitude larger than the average degree. In social networks where the average degree is a good representation of the typical degree of connectivity, it allows to analyse which of the individuals have more or less connections than the typical node. The compactness of a network – or of different of its subgraphs – is an interesting measure to evaluate one aspect of the *group cohesiveness* of a social organisation. While social networks with high compactness exhibit a high level of cohesiveness, they are likely to run into scalability issues as the network grows. In general, large-scale social networks which support efficient information exchange are expected to be sparse, meaning that their compactness is relatively small.

Diameter and average distance. As argued in Sect. 7.2.1, a further important characteristic of a network topology is its *diameter*, which is defined as the longest shortest path between any two nodes in the network. Similarly, the *average distance* gives the average length of shortest paths between any pair of nodes. Both quantities play an important role in the analysis of enterprise social networks, since they quantify how efficient individuals can communicate across shortest paths. A large diameter indicates the presence of at least one pair of individuals that are connected only via a long path. Even worse, a large average distance indicates that the characteristic length of shortest paths between individuals is large. Enterprise social networks supporting efficient information flow between employees are thus expected to exhibit short average distance and diameter.

Measures of connectivity. Capturing the resilience of a network, its node (or edge) connectivity is defined as the amount of nodes (or links) that have to be removed before it falls apart in different components. Both notions of connectivity are illustrated in the network shown in Fig. 7.3, which has a node connectivity of one

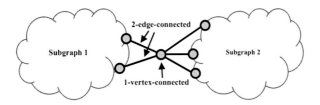

Fig. 7.3 Illustration of the difference between edge-connectivity and vertex-connectivity

and a link connectivity of two (assuming that both subgraphs have higher node and link connectivity). Since each node is connected to a network by at least one link, the link connectivity of a network is always at least as high as its node connectivity. A different approach to quantify the connectivity of a complex network is in terms of its *algebraic connectivity*, a measure which is defined as the second smallest eigenvalue in the spectrum of eigenvalues of a networks' Laplacian matrix.[1] The *algebraic connectivity* can be seen as a generalisation of a network's *connectedness*, where connectedness captures whether all nodes in the network belong to a single connected component. An algebraic connectivity of zero indicates that the network is disconnected, while connected networks exhibit non-zero values. For connected networks, the actual value of algebraic connectivity has been shown to reflect how "well-connected" the network is. In particular, a large algebraic connectivity indicates (a) high node and link connectivity, and (b) small diameter, while small values indicate the opposite [3, 16, 18]. In the context of enterprise social networks, node, link and algebraic connectivity are important approaches to quantify both their resilience and efficiency. Node and link connectivity is crucial for resilient social organisations, since the failure of low connectivity nodes or links can severely impact the network structure, for example, by a separation of communities. The node and link connectivity of networks can be used to identify such critical nodes and links. Furthermore, networks with node and link connectivity exhibit *small cuts* in the topology, which – apart from being a threat to resilience – can be interpreted as bottlenecks that inhibit the diffusion of information. Combining both node and link connectivity and diameter, algebraic connectivity can be used as a measure which jointly captures the efficiency of information flow in a social organisation as well as its resilience: First of all, a large value of algebraic connectivity indicates that all individuals can communicate with each other via short paths. However, it also shows that there are no bottlenecks in the sense that a large fraction of paths necessarily pass through a small set of nodes or links. Algebraic connectivity can thus be interpreted as a measure for the *cohesiveness* of a social organisation.

---

[1] The Laplacian matrix $\mathbf{L}$ of an undirected network is commonly defined as $\mathbf{L} = \mathbf{D} - \mathbf{A}$, where $\mathbf{A}$ is the usual binary adjacency matrix of the network and $\mathbf{D}$ is a diagonal matrix where diagonal elements contain the degree sequence of the network.

7.3 Mining Socio-Technical Systems: Application of Network-Theoretic Measures

The measures introduced in Sect. 7.2.2, as well as their interpretations provided above, highlight interesting opportunities for the monitoring and analysis of enterprise social networks. In practice, data which are suitable to construct and analyse such social networks can come from a variety of social software used in an enterprise context, including social networking tools, collaboration platforms, messaging systems or project management tools. In this section we exemplify this approach using a data set of time-stamped collaborations obtained from a web-based project management tool used by distributed software development teams. In particular – utilising data on Open Source Software communities which have previously been used in the studies [6, 15, 19, 20] – we exemplify some of the metrics introduced in Sect. 7.2.2 and provide a complementary, in-depth analysis of the social organisation of two projects that are the GENTOO project and the ECLIPSE project.

7.3.1 Monitoring Open Source Software Communities

A particularly important and widely used class of enterprise social software that allows to construct social networks are project management tools which support the collaboration, communication or task-allocation in distributed teams. In the context of distributed software development teams, *issue tracking tools* are an important example which allow to report, prioritise and filter reports about software defects, as well as coordinate the efforts to solve them. Such tools are widely used not only within an enterprise scenario, but also in *Open Source Software* (OSS) projects. Since these tools are publicly available to users and contributors of the project, it is possible to extract rich data on the evolving social organisation of these projects. In the following, we thus utilise these data as a proxy to study evolving social structures of humans collaborating on a project. We particularly focus on OSS projects which use BUGZILLA, a popular issue-tracking tool which is widely used in the development of both open source and commercial software projects. While the same data set has been used in [19] to study 14 OSS communities, here we provide detailed results for two major OSS communities: The first is GENTOO a project developing a LINUX-based operating system. The second project is ECLIPSE, which develops and maintains one of the most popular integrated software development environments.

Our approach is based on a construction of *evolving social networks* based on time-stamped interactions between team members that are recorded in the BUGZILLA installation of a project. All recorded interactions within BUGZILLA evolve around *bug reports*, which typically contain a collection of information about a particular software defect. Here, we make use of so-called *Assign* and *CC* interactions, which have a special semantics in the context of issue tracking:

A *CC* interaction between a team member *A* and *B* implies that *A* forwards information about a bug report to team member *B*. An *Assign* interaction between *A* and *B* means that *A* assigns the task of resolving a bug (e.g., by providing a software fix or workaround) to another team member *B*. In the following we take a maximally simple perspective and say that any interaction between *A* and *B* implies that *A* is aware of *B*, thus allowing us to construct a network of collaborating team members. In particular, we consciously sacrifice the additional semantics of different interaction types, as well as their potential implications for the role of individuals, for the sake of simplicity. Since all interaction events recorded in BUGZILLA have precise time stamps, we can further construct *time slices* $[t, t + \delta]$ of social networks by only taking into account interactions happening between time stamps t and $t + \delta$. Using a window size δ of 14 days and an increment $t \rightarrow t'$ of 1 day, we perform a *sliding window analysis*, eventually obtaining a sequence of evolving collaboration networks covering the periods $[t, t + \delta], [t', t' + \delta]$ and so forth. Figure 7.4 shows two example networks constructed from 14 day time slices of the GENTOO project. Nodes in this network represent team members who have been active in the project's BUGZILLA installation within a period of 2 weeks.

Having constructed sequences of collaboration networks for a project allows to apply network-centric measures, thus capturing characteristics of the project's social organisation. Figure 7.5 shows the evolution of six metrics introduced in Sect. 7.2.2 for the project GENTOO. Figure 7.5a, b show the number of nodes and links in the largest connected component of the collaboration networks spanning a period of 2 weeks. One observes significant changes in the number of nodes and edges, highlighting two remarkable phases of growth between December 2003 and February 2006, as well as between June 2010 and April 2012, when our data collection stopped. In addition, a phase during which the number of nodes decreased can be observed between January 2006 and June 2008, followed by a phase of stagnation between June 2008 and June 2010. The number of links representing interactions between team members qualitatively follows the dynamics in the number of active nodes. The number of nodes and links highlights a non-stationary level of team

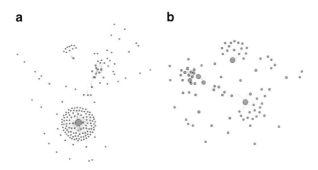

Fig. 7.4 Two collaboration networks of the GENTOO community, constructed from interactions recorded over a period of 14 days in mid 2007 and mid 2009. (**a**) 2007. (**b**) 2009

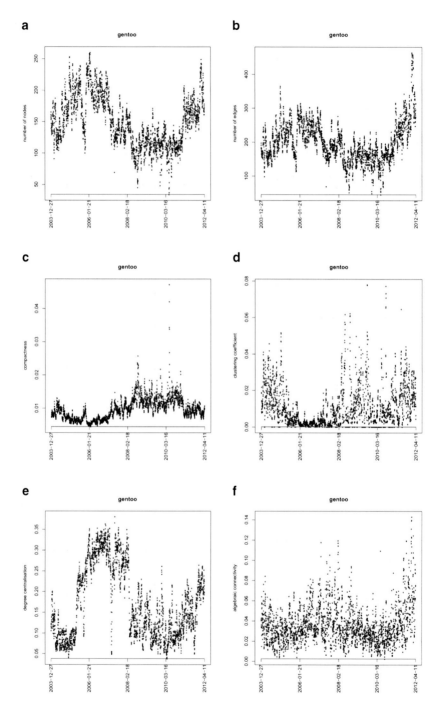

Fig. 7.5 Evolution of network measures capturing social organisation in the GENTOO project.
(**a**) Number of nodes. (**b**) Number of edges. (**c**) Compactness. (**d**) Average clustering coefficient.
(**e**) Degree centralisation. (**f**) Algebraic connectivity

activity and can thus help to interpret the dynamics of other characteristics that are typically affected by the network size. As argued in Sect. 7.2.2, the compactness of a network is a simple size-independent measures which can be interpreted as a particularly simple proxy for the *cohesiveness* of a social organisation. In the GENTOO community, we observe a first phase of decreasing compactness between December 2003 and February 2006, which coincides with the first phase of growth. After a phase of stagnation and moderate increase between February 2006 and March 2008, the compactness of the social network doubled around March 2008, indicating an increase in cohesion. The average clustering coefficient shown in Fig. 7.5d shows a similar dynamics. A first phase lasting until February 2006 shows a remarkable decrease of the average clustering coefficient. During a second phase between February 2006 and March 2008, the average clustering coefficient is remarkably small. The increasing compactness starting in June 2008 was accompanied by an increasing embedding of nodes in densely connected clusters. In Fig. 7.5d, the phase between February 2006 and March 2008 is particularly noteworthy. One can get a clearer picture of the processes shaping the social organisation during this phase by considering additional network-centric metrics. Figure 7.5e shows the evolution of degree centralisation, a measure defined based on the distribution of node degrees. A value of one represents a maximally centralised situation in which all nodes are only connected to a single central node, while a value of zero represents a situation where all degrees of nodes are equal. The degree centralisation shows a remarkable dynamics, exhibiting a highly centralised phased between mid 2005 and March 2008, with centralisation quickly dropping around March 2008. An interview with past and current members of the GENTOO issue tracking team performed in [15] revealed that – between mid 2005 and March 2008 – most of the work associated with the processing of bug reports was done by a *single team member*. Following a dispute with other team members, and being overburdened with tasks, this central member left the project unexpectedly in March 2008. Following this event, the community actively took efforts to reorganise the bug handling process, which is likely to be the reason behind the increasing compactness and clustering coefficient. The evolution of algebraic connectivity depicted in Fig. 7.5f shows a high variability, with a slightly increasing trend between the end of 2005 and the beginning of 2008. Interestingly, the reorganisation of the community following the loss of the central contributor was accompanied by an observable decrease of algebraic connectivity until mid 2010, after which it increased significantly.

Two collaboration networks illustrating the difference in social organisation during the presence of the central contributor between 2005 and 2008, compared to the time after she left are shown in Fig. 7.4a, b. It is tempting to relate the obvious changes in the social organisation discussed above with changes in the performance of the bug handling process during the same period. A study of bug handling performance in the GENTOO community has recently been presented in [15]. It shows that the performance in terms of number of reported/resolved bugs, as well as in terms of the time taken between the submission of a bug report and the first response of a community member, show an interesting dynamics that is likely to be correlated with the evolution of social organisation. In particular, here it was shown the performance of the GENTOO bug handling community generally increased until early 2008. A rapid increase in the response time as well as in the number of open bug reports can be observed at the time when the central contributor left, followed by a phase of stagnation until early 2011 after which performance increased again.

Applying the same measures as above and highlighting differences in the dynamics of social organisation, we now turn our attention to the project ECLIPSE. Figure 7.6 shows the evolution of six network-centric measures over a period of almost 10 years. A first remarkable observation is a pronounced periodicity in the number of nodes and edges, as well as in compactness and the average clustering coefficient. Both the number of nodes and edges in the collaboration network experience a steep increase of up to 500 % roughly once a year. While we cannot make definite statements about the underlying reasons, it is likely that this periodicity is related to the project's release cycle, which aims at one release per year. This increase in activity is associated with increases in both the compactness (Fig. 7.6c) as well as the clustering coefficient (Fig. 7.6d). A further remarkable fact is that – while slight periodic peaks can be observed – compared to the GENTOO project – degree centralisation remains at a rather moderate level also in phases of high activity. One may interpret this as a sign of a *healthy* social organisation, in the sense that an increase of activity is associated with an increase of cohesion, rather than an unproportionate burdening of a few team members.

7.3.2 Analysing Resilience in Online Social Networks

Online social networks are socio-technical systems in which users interact through an online medium, overcoming some of the limitations of verbal face-to-face communication. To improve user experience, the technological component of an online social network is subject to be changed and redesigned, introducing modifications in the fundamental way in which individuals communicate. The impact of such changes in a social system is not trivial, as user reactions are coupled to each other. A technological change, such as a new user interface, can trigger some users to leave the social network, which can decrease the quality of the experience of its friends. This mechanism leads to cascades that can potentially lead large amounts of users away from the social network.

Users leaving a social network can be modelled through a decision process, in which a user receives a benefit and a cost associated with being active in the network. In terms of social interaction, the benefit perceived by a user comes from its contacts, either in the form of information, or as attention. The benefit of a user increases with the amount of active friends, and decrease every time a friend becomes inactive. Costs do not necessarily need to be economic, they can also include the time spent in a social network, or the opportunities lost by not using other platforms. This cost can be increased due to changes in the user interface, or due to service limitations, threatening the cohesion of the network as a society. This model allowed us to analyse the cascades of users leaving the network [5], which stop in subsets of the

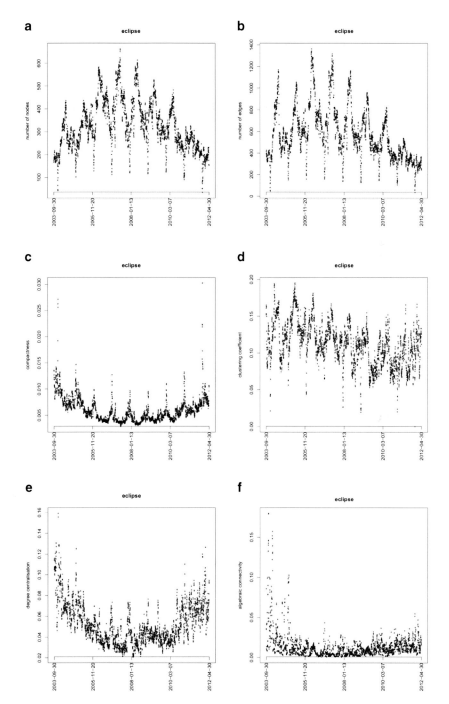

Fig. 7.6 Evolution of network measures capturing social organisation in the ECLIPSE project.
(**a**) Number of nodes. (**b**) Number of edges. (**c**) Compactness. (**d**) Average clustering coefficient.
(**e**) Degree centralisation. (**f**) Algebraic connectivity

network that corresponds to the k-cores explained in Sect. 7.2.1. Thus, the k-core decomposition of a social network allows us to measure its social resilience, i.e. how does the network withstand external shocks and stresses.

We empirically analysed the social resilience of a variety of online social networks, through their k-core decomposition. Such empirical analysis, including large sites like Facebook, MySpace, and Friendster, showed that the topology of these networks can vary a lot in terms of resilience, calling for methods that can increase this desired property. For example, friendship links can be recommended in a way such that links that increase coreness are encouraged, or changes can be introduced gradually to limit cascades of departing users.

These results show that a quantitative analysis of data on enterprise social networks can provide interesting insights into the evolving social organisation of teams, projects or communities. In the case of the GENTOO project, our results show that a monitoring of degree centralisation and average clustering coefficient may have been used as an early indicator for a detrimental evolution of social structures. Furthermore, it is at least conceivable that a targeted optimisation of the network's resilience against the loss of its most central node may have prevented downstream problems with the performance of the bug handling process. In the following chapter, we thus review approaches from the planning and design of telecommunication networks, and discuss their possible application to enterprise social networks.

7.4   Network Planning and Design: Application to Enterprise Social Networks

Analogously to enterprise social networks, the overload or loss of nodes can severely impact telecommunication networks. Therefore, telecommunication providers monitor their networks to allow for an early identification of emerging problems and to take appropriate measures to mitigate their impact on performance. The most important goals of these interventions are to optimise resilience of the network against failing nodes or links, but also to balance load across links and nodes in order to avoid congestion and overload of single nodes which may significantly decrease the performance and efficiency of the entire network – similar to the GENTOO project. Typical interventions include the addition of nodes or links to increase resilience, the rewiring of links or an adaptation of *link capacities* to optimise traffic flows, or the addition of special functionality nodes to manage or monitor regions of a network. In more general terms, *network planning and design* refers to the process of designing the topology of telecommunication networks in a way that optimises some notion of *value*, while keeping *costs* as small as possible. In the following, we will sketch how network optimisation is achieved in general in the context of telecommunication networks. We then we apply it exemplarily to a collaboration network extracted from the GENTOO Open Source community and summarise research challenges arising when wanting to optimise and influence social network structures.

7.4.1 Network Optimisation

In general, the optimisation of network topologies consists of the following three steps: In a first step, an optimisation objective has to be defined based on a notion of *value* defined for particular network topologies. Depending on the context and the associated objective this value can be defined based on different, not necessarily correlated, typically network-centric measures. To give an example, if resilience is to be optimised, one may utilise the average coreness of nodes or the (algebraic) connectivity of the network, while one may chose the average shortest path length, if the latency of communication as to be optimised. In the context of enterprise social networks, the value could, e.g., be defined based on measures capturing communication efficiency, resilience or aspects which influence work atmosphere, thus seeking to balance different aspects by means of a multi-objective optimisation. In general, in the following we assume that the value of a network can be defined as a function of the network topology, which uses the topological structure, node properties, and link weights to quantify the value of the network in a particular context. As second step, an adequate degree of abstraction has to be found to model the telecommunication network for the optimisation process. There is a wide range of different abstraction levels reaching from simple adjacency and distance matrices over partial or complete lists of all possible routing paths up to object oriented representations of each single link and node. The chosen model is one of the essential influence factors for the thirds and final step: the actual optimisation. Probably most optimisation methods known in science and engineering can and have already been applied to network optimisation, including, e.g., (mixed integer) linear programming approaches, different heuristics such as simple greedy mechanisms, simulated annealing, evolutionary algorithms, and – if computationally feasible – the exhaustive evaluation of the entire search space of optimisation options. Each of these methods has advantages and disadvantages, in particular concerning the optimality of the results, the computational complexity of the optimisation, and the capability to handle large network topologies. But they have in common that they aim at maximising the value of the network while balancing it with the associated costs.

7.4.2 Application to an OSS Collaboration Network

In Sect. 7.3, we have seen that a monitoring of network-centric measures capturing resilience (like, e.g., degree centralisation) or aspects that influence communication efficiency (like, e.g., algebraic connectivity) can provide valuable insights into the changing social organisation of software development teams. Furthermore, we have argued above that these aspects are important criteria which are typically accounted for in the design and planning of telecommunication networks. In the following, we briefly describe methods used in network optimisation and discuss their potential applications in the context of enterprise social networks. We further use one snapshot of the collaboration network extracted from the GENTOO community during a period of 4 weeks in May 2002 to illustrate their application in social networks.

Routing Optimisation. To save operational costs the network infrastructure and resources need to be highly utilised without overloading single entities. To cope with the daily dynamics of traffic, resources need to be added in peak hours. This can be achieved by flexible resource allocation and dynamic routing. Although many mechanisms have been investigated to use network resources efficiently, routes in the Internet are still static in many cases. Especially in case of link or node failure, alternative routes that have to carry the traffic of the broken connections are likely overloaded. The aim of routing optimisation is to balance the load on the links of a network [8]. To add resilience, routing is further optimised such that the load is balanced in case of link or node failure. In companies load balancing is important to unburden central employees. It is also reasonable to route, or assign tasks in such a way that information flow is resilient against node or link outages. That means that the workload is still balanced among employees if worker fails. Assuming that each node produces the same amount of tasks, the load on the communication channels can be estimated. To balance the load on the communication channels we consider the 2-core of the communication network, since there exist no alternative links for stub nodes. Stub nodes are nodes that are connected only with one link to the large component to the graph. Figure 7.7a shows the link load in the OSS network if routing is not optimised. The edge colour depicts the utilisation of a link. If routing is not optimised, the links from 22 to 86 and from 22 to 105 are highly loaded, probably resulting in an overload of the involved individuals. Furthermore, if

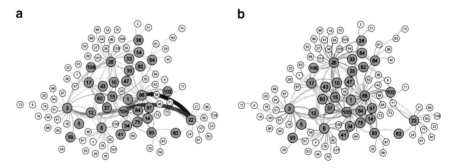

Fig. 7.7 Illustration of different routing layouts in the GENTOO collaboration network. The color and the strength of the edges depicts the load on the link (*red*: high, *yellow*: medium, *blue*: low). (**a**) Unoptimised routing. (**b**) Optimised routing

individual 105 fails all tasks forwarded by node 22 would have to be completed by node 86, which significantly burdens the central contributor. To optimise routing, load is taken from the overloaded links and shifted to alternative routes. In this example load is shifted to node 62. In enterprise social networks, shifting load could be realised by delegating tasks to a different set of workers. Thus, the load is balanced among the paths in the network. If node 105 fails, the tasks originating from node 22 or one of its neighbours, are still shared by node 62 and 86. The load on the links with optimised routing is depicted in Fig. 7.7b.

Controller Placement. In communication networks controllers are needed for authentication, authorisation and connection establishment, but also for dynamic network control adding network functions and rerouting. To monitor network flows efficiently and to be able to access and control all routers in the network efficiently, it is important to place the controllers strategically. In telecommunication networks this process is called controller placement [7]. "Controller placement" can be applied to companies in the sense that workers are chosen as controllers to efficiently spread information or to promote workers which will take a lead in assigning tasks and delegating responsibilities. This type of "promotion" is actually an important mechanism in the bug handling communities of Open Source Software communities, since it is typically only a small set of privileged individuals which is allowed to assign tasks to other community members or developers. If the GENTOO network is considered, and it is assumed that community has the capacity to promote three workers, the question is which workers to choose. Figure 7.8a shows the collaboration network of the GENTOO project. Three controllers – highlighted by a larger node size – are placed in the network in a way that optimises the maximum latency from each node to the nearest controller. Here, nodes 3, 64 and 86 are selected. Node 86 is the central contributor in the GENTOO community and has direct access to a large part of the community. Nodes 3 and 64 are less important, but nevertheless central, nodes covering different parts of the community. Figure 7.8c shows the nodes that are assigned to each of the three controllers. The number of worker assigned to nodes 3 and 64 is small compared with the much larger amount of workers assigned to node 86. That means node 86 would be responsible for many members in the community, which puts a high load on this central contributor. To unburden this central contributor, responsibilities can be delegated to different members, in a way, that each leader is responsible for an equal amount of members in the community. In Fig. 7.8d the controller placement is optimised for lowest load imbalance. Three controllers are placed and associated to a subgraph, such that the load is balanced equally among the leaders. Now, each of the controllers is responsible for 35 or 36 nodes, hence the number of workers assigned to each leader is much more balanced. The drawback is, that the path length and thus the latency on the communication channels between the leaders and their assigned workers can be higher. This can be seen in Fig. 7.8b that shows the latency to nodes for the case of an assignment that optimises load imbalance.

Network planning. Another possibility for the optimisation of communication networks is network planning, i.e. changing the topology of a communication

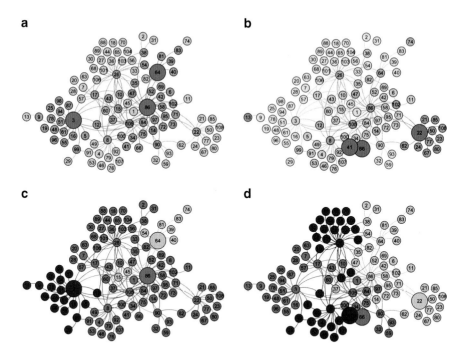

Fig. 7.8 Illustration of best controller placements according to different metrics. (**a**) Lowest maximum latency. (**b**) Lowest load imbalance. (**c**) Assigned members (lowest max. latency). (**d**) Assigned members (lowest load imbalance)

network. While in telecommunication networks, adding nodes can easily be accomplished by setting up hardware, in the context of enterprise social networks this would mean the addition of individuals (i.e. by hiring or transferring between groups). While this is not generally impossible, in the following we focus on targeted interventions by means of *adding links*, which can be achieved in a social organisation in much easier way, e.g. by asking people to collaborate or influencing enterprise social software in a way that it suggests contacts or communication. The addition of links in a social network can bring several improvements, however at the cost of additional expenditures. By adding a link between two persons, i.e., making them direct friends in the network, the path lengths between the existing friends of both persons might decrease leading to a shorter average path length in the network. Another benefit is that additional links can reduce the risk of a disconnection of a person or subnetwork from the rest of the network. A potential increase of the average coreness of nodes can be used as one possible measure to quantify the resulting increase in resilience.

Figure 7.9 exemplarily visualises the improvement reached by adding additional links to the GENTOO collaboration network. In Fig. 7.9a all possible options of adding a single link to the network have been compared regarding the decrease in average path length (right y-axis) and the increase in average coreness (left y-axis).

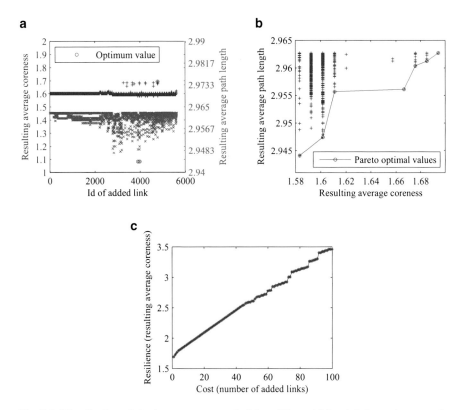

Fig. 7.9 Visualisation of the improvement reached by adding additional links to the network. (**a**) Exhaustive evaluation of all options to add one link. (**b**) Trade-off between two objectives when adding one link. (**c**) Stepwise addition of further links in a greedy manner

The optimum links according to both objectives have been highlighted. Two main findings can be observed: First, there is a high optimisation potential but the improvement reached by adding a link highly depends on the chosen link. Second, the best links to add regarding the optimisation of average path lengths (which can be seen as a proxy for communication efficiency) *do not* correspond to the best ones regarding the average coreness (which can be interpreted as a measure for resilience). In Fig. 7.9b another visualisation is shown that substantiates the latter finding. For all options to add a single link, on the x-axis the average coreness resulting from adding this link is shown, while on the y-axis the resulting average path length is shown. Each symbol in the plot corresponds to one or several links that lead to the same values on x- and y-axis. The highlighted symbols show the Pareto-optimal results regarding both metrics. Looking at Pareto-optimal results is a formal way to identify trade-offs in multi-criteria optimisation.

Finally, to look at the possible improvements when adding several links, a simple experiment has been conducted regarding the average coreness in the network. Subsequently, more and more links were added to the network. In each step of the iteration, all options to add a single link were tried and the option leading to the best improvement was chosen. Figure 7.9c shows for up to 100 added links the reached improvement in terms of average coreness. The linear relationship between resilience and cost highlights a trade-off to be decided by the network provider – or the company when looking at enterprise social networks.

7.4.3   Optimising Enterprise Social Networks: Research Challenges

Realising resilience in enterprise social networks and applying mechanisms from the design and planning of telecommunication networks poses a set of research challenges. These challenges arise from the scale of enterprise social networks, the applicability of the mechanisms, but also from the temporal dynamics of the network topology. In particular, it is not yet well researched how to model enterprise social networks and how to apply the well-known network communication methodology in the domain of enterprise social networks. In this section, several research questions are discussed, whether and how social network analysis may be beneficial to identify and react to problems in the enterprise in advance. Further, we will address the optimisation of an enterprise social network to improve resilience, effectiveness, and job satisfaction. However, beyond the technical aspects, the derived models and applied mechanisms also lead to ethical issues, which we discuss in a separate paragraph.

Modelling Social Systems.   Social network structures in enterprises can be modeled in many ways and an appropriate representation has to be chosen depending on the relations and interactions within the company. For example, links can be created which resemble *boss-of*-relations (i.e. hierarchical structure) within a company or *communicates-with*-relations (i.e. actual communication structures) within the enterprise. Each network will have a specific structure depending on, e.g. enterprise hierarchy or communication flows, some of them are better suited for one company than for another. Thus, the question arises how to model the social network structure which is useful to identify problems or to optimise the network.

   In every company there are persons which are more or less important. How should these personal attributes be modelled and how can their the workload and productivity be quantified? How can team work be integrated into the model? A company does not only need workers who are very targeted and finish one task after another. A company also needs workers that spend time socialising, finding new contacts, and connecting people. Such workers are essential for an efficient working atmosphere. How can their contribution be quantified? Which mixture of personalities is best for the company? Central nodes, for instance, are important to connect persons from different departments and to spread information, but might also suffer from high workload and stress. How can important nodes be identified? How can overloaded workers be detected and unburdened?

If an important person leaves the company, is ill for several weeks, or is moved to a different department, it can have bad consequences for the company like less effectiveness or productivity. What approach can detect and quantify such pending problems? If an enterprise has capacities to employ new workforces, the question is how new nodes are integrated, i.e. to which persons they are linked in the network? Are there other means to change the structure of the enterprise social network? Facilities (e.g. staff rooms) or events (e.g. company outings) support the dialogue of employees which are not directly working together. Thus, new links are created in the social network which foster serendipity and creativity especially when people of different disciplines get together. How can these means be modelled? How can the effect of such means be quantified?

Optimisation Capability. When optimising enterprise social networks with resilience mechanisms new challenges arise from the size of enterprise social networks, the temporal dynamics of the network topologies, but also from the means of modifying the network. Existing communication network heuristics for optimal solutions are typically limited to static networks, which – as has been shown in a recent line of research on temporal networks [9, 12, 13] – can differ significantly from actual communication flows that are due to the temporal ordering of interactions. Moreover, the difference in the structure between communication networks and enterprise social networks makes it difficult to apply common heuristics. Approaches such as routing optimisation, resilience analysis, or network planning are especially efficient in networks where the average node degree and connectedness of the network is already quite high. Furthermore, if personal interests and preferences of the employees are dominant and cannot or must not be influenced, the network structure is fixed. Other mechanisms, such as different message routing, can still improve the resilience of the social network. Instead of just processing plenty of requests, a new working directive could instruct central nodes to forward requests to different communities. Thus, load is taken of the central node and collaboration is enabled between the communities. Finally, compared to telecommunication networks, in social networks it is often more difficult to change or add links or nodes. Here again new challenges arise. Which real-world human resource actions resemble which communication network optimisation? What (side) effects occur when applying such actions in an enterprise? Does the gain of such actions exceed their costs? How do employees' personal attributes change when such actions are enforced?

Ethical Issues. Finally ethical issues have to be considered. It is important to understand that the analysis does not deal with abstract nodes but with humans. Thus, it remains an open issue if a person can be judged by the structure of its social network projection. Moreover, in order to assess personal properties like work capacity or interaction capacity, working behaviour will have to be measured and captured in statistics. However, complex human personalities cannot be fully assessed by such statistics and a supervised working environment will induce changes in attitude and behaviour of workers.

If the social network graph is based on the communications within an online enterprise social network tool (e.g. email communication or Intranet platforms) where network structure and interactions can be easily obtained, different communication or collaboration channels, like face-to-face communication, cannot be captured. The question arises to which extend people are then discriminated who do not extensively use such a tool? Also [14] stated that "in contrast to conventional office software, micro-blogging implies social interaction and self-disclosure. This applies to social software in general. As a result, bringing applications like micro-blogging into the workspace goes beyond traditional technology acceptance theory."

When analysing an enterprise social network, workers might be identified whose working capacity is insufficient. Then, the optimisation of the network can lead to dismissal or demotion of workers. Moreover, forced human resource actions like moving persons to another department might have complex impacts on the employees' motivation. Also other resilience means might encounter employees' resistance. For example, the expected establishing of connections between different workers might not be accepted as it overrides the workers' own preferences for selecting social peers. Thus, it remains an open question which actions can – and should – be enforced by the company without running into ethical issues.

7.5   Conclusion

The resilience and efficiency of communication networks is a major topic in the network research communities both studying social and telecommunication networks. With the rise of collaboration platforms in enterprises social network structures on top of technical systems emerge which reflect the social organisation of an enterprise. Therefore it is tempting to utilise known results and insights from the optimisation of telecommunication networks. This maybe helpful for enterprises to improve their human resource management by pre-emptively taking load of busy and central workers and improve the social network structure to increase information diffusion and to accomplish a healthy work environment. The main contributions of this chapter are (a) to summarise network-theoretic measures and interpret their meaning in the context of enterprise social networks, (b) to illustrate how enterprise social networks can be monitored by showing an example from OSS communities, and (c) to demonstrate approaches from the optimisation of telecommunication networks and to apply them to a real-world collaboration network.

Our results highlight new technical, scientific and ethical challenges which arise when wanting to monitor and optimise enterprise social networks. Combining expertise from the modeling and analysis of complex networks, the design and optimisation of telecommunication networks, as well as from the social sciences, the emerging interdisciplinary field of socio-informatics has the potential to address these challenges.

References

1. Cohen R, Erez K, Ben Avraham D, Havlin S (2001) Breakdown of the internet under intentional attack. Phys Rev Lett 86:3682–3685. doi:10.1103/PhysRevLett.86.3682, http://link.aps.org/doi/10.1103/PhysRevLett.86.3682

2. Dunbar R (1998) Grooming, gossip, and the evolution of language. Harvard University Press, Cambridge

3. Fiedler M (1973) Algebraic connectivity of graphs. Czechoslov Math J 23(2):298–305

4. Freeman LC (1977) A set of measures of centrality based on betweenness. Sociometry 40(1):35–41. http://www.jstor.org/stable/3033543

5. Garcia D, Mavrodiev P, Schweitzer F (2013) Social resilience in online communities: the autopsy of friendster. In: Proceedings of the first ACM conference on online social networks (COSN'13), Boston. ACM, New York, pp 39–50. doi:10.1145/2512938.2512946. http://doi.acm.org/10.1145/2512938.2512946.

6. Garcia D, Zanetti MS, Schweitzer F (2013) The role of emotions in contributors activity: a case study of the GENTOO community. In: Proceedings of the 3rd international conference on social computing and its applications (SCA2013), Anaheim

7. Hock D, Hartmann M, Gebert S, Jarschel M, Zinner T, Tran-Gia P (2013) Pareto-optimal resilient controller placement in SDN-based core networks. In: 25th international teletraffic congress (ITC), Shanghai

8. Hock D, Hartmann M, Menth M, Schwartz C (2010) Optimizing unique shortest paths for resilient routing and fast reroute in IP-based networks. In: IEEE/IFIP network operations and management symposium (NOMS), Osaka

9. Holme P, Saramäki J (2012) Temporal networks. Phys Rep 519(3):97–125. doi:http://dx.doi.org/10.1016/j.physrep.2012.03.001, http://www.sciencedirect.com/science/article/pii/S0370157312000841

10. Newman ME (2010) Networks: an introduction. Oxford University Press, Oxford/New York

11. Olguin D, Waber B, Kim T, Mohan A, Ara K, Pentland A (2009) Sensible organizations: technology and methodology for automatically measuring organizational behavior. IEEE Trans Syst Man Cybern Part B Cybern 39(1):43–55. doi:10.1109/TSMCB.2008.2006638

12. Pfitzner R, Scholtes I, Garas A, Tessone CJ, Schweitzer F (2013) Betweenness preference: quantifying correlations in the topological dynamics of temporal networks. Phys Rev Lett 110:198,701. doi:10.1103/PhysRevLett.110.198701, http://link.aps.org/doi/10.1103/PhysRevLett.110.198701

13. Scholtes I, Wider N, Pfitzner R, Garas A, Juan Tessone C, Schweitzer F (2013) Slow-down vs. speed-up of diffusion in non-Markovian temporal networks. ArXiv e-prints. http://arxiv.org/abs/1307.4030

14. Schöndienst V, Krasnova H, Günther O, Riehle D (2011) Micro-blogging adoption in the enterprise: an empirical analysis. In: Wirtschaftsinformatik, Zurich, p 22

15. Zanetti MS, Scholtes I, Tessone CJ, Schweitzer F (2013) The rise and fall of a central contributor: dynamics of social organization and performance in the GENTOO community. In: Proceedings of the 6th international workshop on cooperative and human aspects of software engineering (CHASE 2013) held at ICSE 2013, San Francisco

16. Van Mieghem P (2012) Graph spectra for complex networks. Cambridge University Press, Cambridge/New York
17. Wasserman S, Faust K (1994) Social network analysis: methods and applications. Structural analysis in the social sciences. Cambridge University Press, Cambridge/New York
18. Wu CW (2005) Algebraic connectivity of directed graphs. Linear Multilinear Algebra 53(3):203–223
19. Zanetti MS, Sarigöl E, Scholtes I, Tessone CJ, Schweitzer F (2012) A quantitative study of social organisation in open source software communities. In: Jones AV (ed) 2012 Imperial College computing student workshop, openaccess series in informatics (OASIcs), London, vol 28. Schloss Dagstuhl–Leibniz-Zentrum fuer Informatik, Dagstuhl, pp 116–122. doi:http://dx.doi.org/10.4230/OASIcs.ICCSW.2012.116, http://drops.dagstuhl.de/opus/volltexte/2012/3774
20. Zanetti MS, Scholtes I, Tessone CJ, Schweitzer F (2013) Categorizing bugs with social networks: a case study on four open source software communities. In: 35th international conference on software engineering (ICSE'13), San Francisco, 18–26 May 2013, pp 1032–1041

Chapter 8
Assessing the Structural Fluidity of Virtual Organizations and Its Effects

Sean P. Goggins and Giuseppe Valetto

Abstract A major advantage of Virtual Organizations (VOs) is flexible membership and participation. VO members are able to join and leave VOs at will, and can change whom they collaborate with at any point in time. Such flexibility may make VOs more efficient in the completion of collaborative work than traditional organizations. However, efficiency is only one of several measures of organizational performance; and flexibility in a virtual organization includes both how VO structures may be more fluid and adaptive, and how VO leadership emerges and evolves throughout the VO lifecycle. The aim of this chapter is to: (1) define and quantitatively assess the actual flexibility of participation in VOs, through a social network index that we call structural *fluidity*; and (2) measure the relationship between fluidity and performance in the work carried out within the VO. These are essential insights for the development of theories to guide the formation, development and dissolution of VOs, and teams that emerge around VO work. To accomplish these aims, we will apply a methodological approach and ontology for the study of VOs that we have used in over a dozen published studies, and refer to as Group Informatics. Our approach enables a comprehensive, interdisciplinary inquiry into the relationship between structural fluidity and performance in diverse VOs. Specifically, we will examine VOs in software engineering, disaster relief, online learning and public discourse communities that emerge through social media. We will apply Group Informatics to the design, development and testing of empirically and theoretically grounded algorithms for measuring VO fluidity and performance in each context, which will result in new theoretical advances that enable sophisticated analysis of the resulting data.

S.P. Goggins (✉)
University of Missouri, Columbia, MO, USA
e-mail: gogginsS@missouri.edu

G. Valetto
Fondazione Bruno Kessler, Trento, Italy
e-mail: valetto@fbk.eu

© Springer International Publishing Switzerland 2014
K. Zweig et al. (eds.), *Socioinformatics - The Social Impact of Interactions between Humans and IT*, Springer Proceedings in Complexity,
DOI 10.1007/978-3-319-09378-9_8

8.1 Introduction and Motivation

One of the contexts of human activity that has been most deeply impacted by the prevalence and ubiquitous presence of Information and Communication Technology (ICT) has been without doubt the workplace. This is hardly new or surprising, since it had been predicted for decades that "office automation" would represent one of the first areas where computerized tools would fulfill their potential, and would revolutionize habits and practices.

Besides bringing along tremendous gains in individual productivity, the ICT revolution has also profoundly changed the concept of the office as an *organizational locus*. It has introduced ever-increasing degrees and sophistication of Computer-Supported Cooperative Work (CSCW); with that it has enabled the breaking of the physical, geographic and firm boundaries, and has helped construct novel technology-supported contexts for coordination and teamwork, which can be created ad hoc to help collaborators to get together and get things done. This is the promise of *Virtual Organizations* (VOs), and, in recent years, we have witnessed the rise of VOs in a large number of different domains, ranging from software engineering to education, and disaster response.

Virtual Organizations (VOs) have flexible membership and participation, as the barrier for members to be able to join and leave a VO is very low (in many cases they can do so at will); moreover, members can adapt with respect to whom they collaborate with at any point in time [52, 37]. This potential for flexibility is regarded as an intrinsic characteristic of VOs, and a differentiating factor vis-à-vis more traditional types of organizations. From such flexibility should descend that the organizational structure that can be observed in a VO can be more fluid than that of a traditional organization.

With this chapter we define the concept of *structural fluidity* of a VO in a way that can be measured quantitatively. Intuitively, structural fluidity is an easy concept to grasp, as an indication of participation, role and leadership changes within the organization over time. The reason why structural fluidity is an important concept to quantify and investigate lies in one of the basic assumptions underlying the interest in VOs, that is, that the flexibility implied by higher degrees of fluidity is an advantage. More specifically, one major expectation is that the greater structural fluidity in VOs is accompanied by higher performance in the completion of analogous work, when compared to more traditional forms of organization [4, 55, 58]. This extends from established understanding of the relationship between higher organizational adaptability and higher likelihood for an organization to survive over time [10].

However, the relation between indicators of structural fluidity and organizational performance has not yet been carefully examined, or proved. In fact, categorical and static qualifications of "how virtual" a VO is are discussed in the literature (based on a composite of attributes such as time zone difference, geographic dispersion, culture and work practice differences), but are **not** sufficient, *per se*, to explain variations in VO performance [11].

Our research agenda on the topic of structural fluidity is articulated according to a twofold objective: (1) develop methods to quantitatively assess the flexibility of VOs by means of the definition of an index of the structural fluidity of an organization. (2) investigate relationships between said fluidity index and context-specific performance indicators for the work carried out within such VOs.

A quantitative index of structural fluidity could shed light on whether VOs tend to be indeed more flexible than comparable traditional organizations, and to differentiate among diverse VOs designs and categories, in terms of their observable flexibility in allowing and accommodating structural changes. Furthermore a structural fluidity index would enable us to observe whether there are significant relationships between structural fluidity and performance. That is the ultimate goal of this kind of research: performance analysis is one principal application of a fully developed theory of Virtual Organizations [59], since it would provide a much needed insight to further the development of theories, practices and tools to guide and support the life cycle of VOs throughout their formation, development and dissolution.

8.2 Approach Outline

In this Section, we briefly introduce the challenges of investigating structural fluidity, and outline how they can be addressed.

We maintain that a foundation for the investigation of structural fluidity is the analysis of electronic traces that are made available by VOs by their very nature, since they can be captured and persisted from the collaborative tools that enable VO members' behavior. Studies that leverage those traces abound; among them, [9, 23, 26, 27, 32, 40, 57] offer some demonstrative examples of analyses that can be applied – and of insights that can be gleaned – which are similar to some techniques we envision to investigate structural fluidity.

Whereas, as we discuss further below, trace-based measures of performance tend to be context-specific and must be adapted to the work domain of each VO, we conceptualize a structural fluidity index as a composition of four general characteristics, which can be all made available from *the social network* fabric of a VO. One major characteristic we identify is leadership and, more precisely, **changes in leadership**, since at specific junctures a flexible VO can promote the rise of different leaders in different organizational positions, and exercising different types of leadership. Furthermore, the measurement of fluidity will also include **macro-properties of the VO** and its social network, including size, group count, and membership volatility. We will also take into account **trends over time of those macro-properties**, for instance, the participation trajectory of the VO, since a VO that tends to accommodate more and more new members is likely to provide for greater opportunities of diverse and emergent collaborations. The last characteristic denoting VO fluidity is a micro-property which becomes evident at the local level in the network, that is, the **variability of the ego-network** of each VO member at different moment in times, since we postulate that a high level of fluidity of the VO results in more free movement of its members in between and within smaller groups (both formal and informal), which in turn effects the variability of the members' "neighborhood".

Although all of the above properties can be analyzed by leveraging electronic trace data of interaction and work, there are several challenges associated to that kind of socio-technical analysis. First, the electronic trace data alone is not usually a complete record of participant interactions [34, 28, 29, 30]. Second, the relationship between these traces and performance requires systematic evaluation [1]. Third, organizational flexibility as measured through the fluidity of the social networks detectable from electronic trace data is difficult to ground both theoretically and empirically solely in analysis of those traces [34, 28, 29, 30]. Fourth, although features, like leadership and leadership changes, can easily be extracted from the social network of a VO, what organizational relationships should be mapped in the social network, and how, may vary significantly across domains and contexts, as shown in many field studies, such as [9, 25, 23, 31, 36, 3, 28, 29, 30]; correspondingly it is not obvious how each network relates to both structural fluidity and performance. Fifth, although the characteristics that we have identified above as contributing to a structural fluidity index can be – individually – relatively easily computed and analyzed, the extent to which each needs to be considered and its "weight" in such an integrated index is not necessarily self-evident; in fact, it may vary depending on the specificity of an individual organization or an organization domain, which calls for contextualization of the index-building procedure.

For all of those reasons new methodological approaches to VO research are required to address these challenges. We have started to address the challenges listed above with a method and ontology for the study of virtual organizations that, which we refer to as *"Group Informatics"* [28, 29, 30]. A principal tenet of Group Informatics is the focus on the small group as the unit of analysis in the field, in recognition of the central role that small groups play in organizational change [33, 35], societal change [19], and ICT use [53, 24]. Another tenet – as we have already mentioned – is the rooting of the analysis in the electronic traces of the interactions that are mediated by the computerized environment of the VO. However, an additional factor is the contextualization of traces with respect to the socio-technical properties of that environment, such as artifact types, members' characteristics, or interaction attributes and meta-data. The contextualization step is important in recognition of the fact that VOs are not a single, uniform construct, but include a set of organization types, which exhibit varying degrees of, for instance, virtuality, stability, and expected duration [11]. For example a VO like Wikipedia exhibits highly formalized structures and processes [6, 39], whereas the long tail of VOs typically lack that level of formality. Finally, Group Informatics seeks the integration of quantitative and qualitative methods of analysis, to allow for triangulation of findings and thus augment explanatory power. A thorough discussion of the Group Informatics approach and how it can shape the investigation of structural fluidity is offered further below, in Sect. 8.6.

8.3 Related Work

8.3.1 Structural Metrics for Virtual Organizations

To work toward closing gaps in our understanding of structural change in VOs over time, we synthesized the following four core measures of organizational change from the literature: (1) VO leadership (degree and betweenness centrality); (2) VO membership (who is participating); (3) VO subgroups (what groups 'move together'); and (4) Changes in VO organization size and structure (network level statistics). To understand the fluidity of a VO, it is necessary to analyze changes in each of these factors over time, and to work toward a synthesis of these factors into an overall indicator of structural fluidity.

Leadership is an individual measure. Understanding how leadership evolves in a VO is a key pathway for understanding structural fluidity. This includes measurement of VO leadership, made through social network analysis (SNA) from two well-established and complementary perspectives. The first is degree centrality, which is a measure that identifies people in the center of the action by counting the connections people have with others. In directed network analysis, connections in and connections out are measured separately and referred to as "in degree" and "out degree" centrality. Central people are either the formal heads of an organization, or central players in the informal organization; people regarded by their peers as possessing attributes that lead them to be referred to a lot. The second way leadership is identified through SNA is through a statistic referred to as betweenness centrality [21]. There are a number of derivatives of this statistic that have been used in different contexts [8, 20], and this measure is incorporated as a component of a methodological approach designed to facilitate valid network analysis from electronic trace data [26].

Betweenness identifies people who sit between groups in a larger organizational context. In a software VO, these are the people who facilitate the integration of code from multiple software teams, or cross a range of topics in online discussions [3, 23, 43]. Betweenness centrality is also referred to as brokerage, indicating that a person is a mediator between two other groups or categories of people in a transaction.

There are two main measures of structural change at the organization level that are used in prior literature on distributed work. First, there are changes to the size and composition of the social connections within an organization [47–49]. Second, a number of studies of VOs in OSS examine social network constructs of core membership, periphery membership and overall network centralization ([14–16]; Kevin [13]). Prior work does not, however, examine these measures of organizational structure over time.

8.3.2 Measuring Change in an Organization Structure

Structural fluidity is conceptually similar to change detection in virtual networks. Change detection research focuses on identifying change in social networks composed either entirely of computational agents in the case of simulations or entirely of real people in the case of applied studies. This work contributes to our understanding of the differences in these types of networks and limitations of existing network analytic techniques for detecting change. McCulloh [44] defined a set of statistical control charts capable of detecting statistically significant changes in social networks. Control charts demonstrate validity in controlled, software agent networks, but have yet to be demonstrated as valid in the analysis of change in social networks involving actors in physical or virtual organizations composed of humans. The challenge with detecting change in human networks is that these networks change a lot; therefore, finding actionable, meaningful changes with computation alone becomes difficult.

To mitigate these challenges, McCulloh's [44] work, and the work of others examining longitudinal statistics for social network evolution are primarily focused on highly structured organizations like the military. In these types of organizations, comparisons across smaller social aggregations may provide immediate information about changes in leadership structure in a platoon. To build understanding of how smaller, decentralized social groups (with no obvious or formal leadership) interact through technology requires research methods that reflexively analyze and triangulate trace data with findings from content analysis, interviews and other qualitative methods [23].

Projects on GitHub provide a new, data rich site for examining organizational change where there is not an a priori structure like one finds in military organizations. VOs on GitHub are emergent virtual and decentralized organizations that generate electronic trace data reflecting a significant portion of the social, task and information behaviors of participants.

Other studies describe organizational practices qualitatively, or examine trace data computationally. For example, Geiger and Ribes [22] propose trace ethnography as one possible methodological approach, but, like Stahl's [53] extensive ethno-methodologically informed analysis of electronic trace data shows, the approach does not scale to large conversations or longitudinal studies of VOs.

In contrast, computational analysis of trace data is demonstrated to be effective for identifying clusters of interaction or keywords from large corpora of data [39]. The main critique of computationally focused analysis of trace data is that often it makes a limited account of social science theories and is not triangulated with data describing the underlying social phenomena. Livne, Simmons, Adar and Adamic [42], for example, contrast network and linguistic analysis of Twitter data during an election cycle in the US, showing the network model as nominally more predictive of the outcome than the language model. Both models, however, are built around a binary choice between two candidates, and were only nominally more predictive than simple selection of an incumbent (88 % for the network model vs. 81 % for the incumbent status model). These and other computational social science studies that do not fully incorporate social science theory in their framing [12, 56] make a more limited contribution to our understanding of how behavioral data about people can be used to describe organizational change.

To measure structural fluidity, some combination of methods from social science [22, 26, 38, 53], prior studies in a particular context [14] and computational methods (Goggins, Valetto, Mascaro, & Blincoe, Published Online First; [46]) are required. An integrated approach will overcome known issues of validity and theoretical coherence associated with the computational analysis of electronic trace data [34].

8.3.3 Measuring the Performance of a Virtual Organization

Measuring performance is contextual, since performance is necessarily defined in a domain-specific way. Consequently, it is hardly possible to come up with generalizations that can be applied to Virtual Organizations within diverse domains. However, having ways to measure performance is of course a paramount concern for those VOs that structure themselves and operate as communities of practice [60], or – according to [5] – network of practices.

We review hereby domain-specific performance indicators in Software Engineering, Online Learning, Disaster Relief and Social Media, which are among the most often studied fields where VOs are deployed and operate. Those are also the fields in which our own interest lies, in terms of experimental and fieldwork aimed at establishing, understanding and structural fluidity assessment vis-à-vis VO performance.

8.4 Software Engineering

Concepts of performance in a software development organization are generally tied to either the quality of the product, or the effectiveness of the process. For our purposes, many product-derived performance metrics are scarcely actionable, as they become available *post hoc*, or at least out of band. For example, the number of residual defects is a major quality factor for a software release, but it is only known after exhaustive in-house testing, or following post-release customer feedback; that is the reason why software engineering research has spent a lot of effort on predictive models that attempt to proxy and anticipate the actual defects, their count, or their density within specific modules of the software product [2, 63]. Other often-used indicators that are used for instance in a prominent VO model like Open Source Software (OSS) projects include adoption [51], or maturity [18, 61] of the software product. The extraction of metrics for all of these indicators typically requires a long observation period.

Metrics that predicate upon the effectiveness of the software process encompass aspects like the organizational health or efficiency. Health metrics are, again, used extensively to analyze OSS. They try to conceptualize the performance of an open source community as its ability to thrive and attract a continued influx of contributions (and contributors), and focus on the numbers of participant in the VO in different recognizable roles ad their trends over time [18], or look at characteristics and patterns within the social network of the VO [13, 15, 62]. Efficiency metrics look at the issue of productivity by evaluating the ability of the VO members to fulfill project tasks quickly and correctly. They originate from a modality of work organization and assignment that has become prevalent in large and distributed software development organizations, in which the atomic units of work are *Change Request* (CRs), which are posted in a public, computer-mediated place (among the most popular tools implementing those CR repositories there are Bugzilla, Jira and GitHub) either by members of the project or external actors (for example, end users of the software). CRs are collectively triaged by the VO for relevance and priority, and then assigned to specific VO members, or self-assigned. The literature – see for instance [9, 45, 65, 64] – proposes a variety of efficiency indicators that can be derived from the timeline and workflow of the CRs that enter the system, including the rate of CR resolution over time; the turnaround time of CRs; the number of CRs that remain open or unassigned; the rate at which code is contributed to resolve open CRs; or the number of such code contributions that get accepted and incorporated in the project code base vs. the number of contribution that do not pass quality assurance and are rejected.

8.5   Disaster Relief

Disaster response is a different specific context than software engineering, but has analogous concerns with performance, process and coordination behavior. Measurement of performance in the use of Internet during a disaster is not yet prevalent, but the research to date, including our own [28], demonstrates that information quality and the presence of coordination behavior are two important factors that contribute to the usefulness of these media during a crisis. This extends Palen et al.'s [50] vision for the future of disaster management, which leverages the use of ICTs by focusing on the potential of members of the public during disaster situations. They suggest that supporting the public and enhancing their ability to make good, timely decisions can reframe disaster relief as a socially distributed information system [50].

Bui et al. [7] developed a framework for conceptualizing the types of issues that emergency relief workers must overcome, suggesting that the central issues in disaster relief management are information, coordination and the effects of disaster relief work on workers [7]. Information issues include information distortion and inconsistencies that must be reconciled. Coordination among governments and NGOs can be problematic due to government reluctance in releasing information with potential security implications and communication incompatibilities, including both language and technology [7].

Measurement of performance in the use of Internet during a disaster is not yet prevalent. However, research to date, including a framework developed by Bui et al. [7] as well as our own investigations [28, 29, 30], reveal that information quality and the presence of coordination behavior act as important factors that contribute to the usefulness of these media during a crisis. Information quality issues include distortions and inconsistencies that must be reconciled, and lack of effective coordination behaviors among governments and NGOs can present problems due to communication incompatibilities, including both language and technology [7], and governmental reluctance to release information with potential security implications. These identified factors extend Palen et al.'s [50] vision for the future of disaster management, which leverages the use of ICTs by focusing on the potential of members of the public during disaster situations. They suggest that supporting the public and enhancing their ability to make good, timely decisions can reframe disaster relief as a socially distributed information system [50].

Bui et al. [7] developed a framework to conceptualize the categories of issues that confront disaster relief workers, suggesting that the central issues in disaster relief management are information, coordination behavior and the effects of disaster relief work on workers [7].

8.6 Methodology

One component of our work to date is the development of a methodology and ontology for conducting research involving electronic trace data. Performance and structural fluidity are our specific concerns in this chapter. To address these questions the more general problem of ensuring a connection between existing theory, research questions and the structure and meaning of electronic trace data is important for ensuring the validity of research [34]. Group Informatics proposes a systematic approach to ensuring a deliberate connection between trace data, theory and research questions. Connecting requires reflexivity between the human review of how individuals, teams and organizations are functioning and computational approaches to the trace data they leave behind in the technologies.

8.6.1 Overview of the Group Informatics Method

With Group Informatics, we have developed a comprehensive methodological approach and ontology for the study of virtual organizations [28, 29, 30]. The concept of interaction is central to Group Informatics, but we are interested in the contextualizing of member interactions, by operationalizing Dourish's view of context as a dynamic construct [17]. Therefore, our approach calls for the

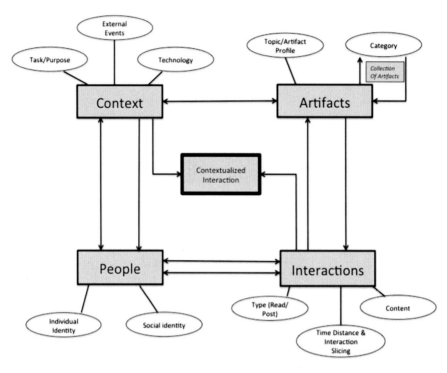

Fig. 8.1 Model overview of group informatics

contextualization, aggregation and weighting of member interactions, according to an ontological model that comprises four core components: (a) artifacts, (b) interactions, (c) context, and (d) people. Each of those components has dimensions and relationships with the other components.

The four components in Fig. 8.1 contribute to into a network of contextualized interactions, which are weighted and can be decorated with additional attributes (meta-data), and which are bound to vary over time. Contextualized interactions may be either interactions between members, or interaction of members with artifacts, in which case artifacts are regarded as boundary objects around which interactions occur [54, 41]. The network of contextualized interactions represents social phenomena within the VO, and we can ask research questions of them and investigate them by means of network analysis methods; in [26, 27], we exemplify the analysis approach enabled by the Group Informatics model, and show how it applies to the question of identifying emergent and informal groups within larger VOs in two diverse cases from the software engineering and online learning domains.

Figure 8.2 conceptualizes how interactions are aggregated and weighted. The aggregations and weights are not computational choices; they are choices based on developing a qualitative, grounded understanding of how people are interacting with

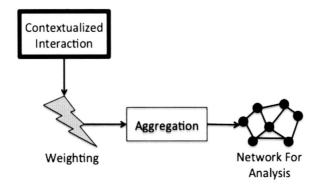

Fig. 8.2 Model for context adaptivity derived from electronic trace data of interactions

each other in the system that generates trace data. Such an understanding emerges from interviews, surveys, and systematic, ethnographic observations of online teams and organizations, as described in Goggins et al. [26, 27].

8.6.2 Conceptualizing and Measuring Structural Fluidity

Structural fluidity is a measure of organizational change. As the dynamics of a VO are self-organized (or at least originated from a mix of directed and self-organized changes), we want to leverage the methodological approach of Group Informatics and its contextualized networks to first measure multiple facets of structural fluidity, and then compose them in a single statistical index.

Some motivations for having a structural fluidity index are: to assess analytically that an organization is evolving; to what extent (how much) it is evolving; and what the trend of this evolution is over time. Our ultimate goal, however, goes beyond observation, and aims to answer the following research question: **to what extent does structural fluidity correspond with performance in VOs?** For that, we want to associate this measure of organizational change to measures of performance across different kinds of VOs.

To attack the problem, we use contextualized networks and apply on them network-analytical metrics like degree centrality. Network statistics capture primary facets of an organization structure and its dynamism; among them, measures of leadership, membership volatility, the size, group count and participation trajectory of the VO. The extent of movement of members between and within qualitatively and quantitatively identified groups, which are sometimes referred to as neighborhoods in Social Network Analysis. Representing the structural dynamism of virtual organizations is then connected to studies seeking to understand how contributors and leaders measure performance.

8.7   An Example from Software Engineering on GitHub

Using prior studies of OSS and organizations, we proposed and evaluated a set of factors to measure the structural fluidity of VOs, and operationalized those factors in a case study of one GitHub project. We identified the *type of work performed* as a potential covariant for understanding structural fluidity in virtual software organizations more comprehensively. The role of the type of work warrants ongoing study.

We demonstrate that the rails/rails project on GitHub has dynamic, distributed leadership; fluid membership; non-static subgroups, both measured from an individual viewpoint (ego networks) and at the VO aggregate level; and variations in the rate of new participation. Together, these factors suggest a type of structural fluidity in the rails/rails organization that is not previously operationalized in open source software development or organization studies. Our fine-grained analysis of contribution type illustrates that individuals take on different work during different time periods.

Our findings lead us to consider three specific areas for continued empirical study of virtual organizations and the development of VO theory. First, our observations offer a starting point for the development of an index of VO structural fluidity, conducive to comparative studies of VOs. Second, we suggest considering how VOs that demonstrate a high degree of structural fluidity may be thought of more like "*impromptu* collaborations" than traditional organizations. Third, we argue that, unlike non-virtual organizations that employ traditional knowledge from management to effectively scale up, VOs present an opportunity for scalable, innovative organizations to embrace an approach more influenced by values of anarchy than hierarchy.

8.8   Quantifying Structural Fluidity

The design of collaborative computing systems is a recognized wicked problem [22]. Understanding the uptake and use of context specific technologies and practices is similarly challenging. Working toward the development of techniques that can offer a comprehensive, comparative measure of VO change is therefore useful for designers and VO stakeholders.

The feasibility and importance of developing a synthetic index of structural fluidity emerges from our findings. We proposed a set of factors that can signify structural fluidity and that are associated with quantitative measures that can be observed directly from qualitative analysis of trace data. Some of those factors are "macro", i.e., regard the organization as a whole: these include the total number of participants (network size), the number of newcomers, and changes in subgroup composition. Other factors are "micro", i.e., regard individual VO actors, and characteristics of the corresponding network position: these include measures of degree and betweenness centrality, and measures of dissimilarity of the ego network of an actor. Both macro- and micro- measures can be observed repeatedly over time, so that it is possible to construct a set of time series (one for each of the factors considered).

In a socio-technical system like a VO, it is unlikely that the values obtained from those repeated observations are independent. One way to conceptualize fluidity is that – although an observation of a factor at any given time is not independent from the accumulated history of the subject (either an actor in the organization, or the organization at large) – the time series of the observations taken as a whole should not show significant patterns, including stability, trends or periodicity. Rather, the more fluid the structure of the organization, the more random the time series should look. Our Ljung-Box tests illustrate the type of randomness expected.

This is consistent with the idea that in a highly fluid VO collaborations are *impromptu*, and past collaborations may not repeat in the future, and do not necessarily dictate how future collaborations shape up. We hypothesize that routine organizational change is a premise for many VOs. Measuring that change and drawing comparisons then becomes essential. One method to assess whether the VO being studied shows fluidity is the use of statistical tests of randomness for the corresponding time series. For instance, the Ljung–Box test can be used to refute the null hypothesis of randomness; its Q statistics measures, so to speak, the "lack of randomness" of a time series, with higher Q values (when coupled with significant p levels) meaning that the time series is further away from randomness. In our terms, though, high Q values signify a lack structural fluidity. We do indeed observe such low structural fluidity among a small set of contributors in sustained leadership and coordination positions. However, most of the organization is highly fluid.

By assessing the randomness (or lack thereof) of the time series for each of the factors we have proposed and explored in this paper, we quantify how much a VO is structurally fluid "according" to that factor. This creates a multi-dimensional criterion for assessing the fluidity of VOs.

Moving from these multiple dimensions towards a unified indicator of the structural fluidity of organizations will require further work to discern which factors provide the most reliable and valid indicators of structural fluidity. We should also investigate what relationships may exist between the various factors, and their measures. There is ample space for further research in this area; for instance, our observations of differences and variability of type of work in GitHub, once fully developed, could become a key for validating the explanatory power of each factor, and the relationships between those factors. We regard our current work as a starting point for examining structural fluidity in individual VOs, and for comparing VOs.

8.9 Impromptu Collaborations: A Path to Theories of Structural Fluidity

Rails/rails has nominal central control in the "merger" role, but a growing number of contributors emerging through GitHub's pull request process. These widespread, diverse collaborations are much more spontaneous, ad hoc, and at times short-lived than traditional organization forms or bureaucratic organization forms like those found in Wikipedia. We suggest that it is possible that GitHub appears to support an organizational model that is neither hierarchical nor tribal in its form.

Rails/rails exhibits a small set of people in hierarchical leadership roles – called mergers – who do the work of building and distributing code (the product). Beyond contributors in those narrow roles, leadership is highly fluid. We do not observe hierarchy; and the volume of productive work makes managed anarchy seem implausible, yet not wholly inconsistent with what our data illustrates. The development of better VO theories will, we think, result from examination of socio-technical environments like GitHub, and openness to a range of unconventional, post-organizational research questions. Structural fluidity, applied as an index across VOs, has a potential to demonstrate its value in this kind of research, in terms descriptive utility.

We present these findings recognizing important limitations and insights. With regards to limitations, the focus on a single VO in GitHub is not generalizable to other VOs, though our ongoing studies suggest rails/rails is similar to many GitHub VOs, just on a smaller scale. This is an exploratory, proof of concept examination of the idea of structural fluidity that lays groundwork for the development of reliable and valid measures of differences in VOs. We show evidence of structural fluidity and explain the role of different types of leadership across a number of indicators that lead us to propose an index to support the ongoing study and measurement of VOs.

References

1. Adar E, Ré C (2007) Managing uncertainty in social networks. IEEE Data Eng Bull 30(2):15–22
2. Arisholm E, Briand LC, Johannessen EB (2010) A systematic and comprehensive investigation of methods to build and evaluate fault prediction models. J Syst Softw 83(1):2–17
3. Blincoe K, Valetto G, Goggins S (2012) Leveraging task contexts for managing developers' coordination. In: Proceedings of the 2012 ACM CSCW conference, ACM, New York, pp 1351–1360
4. Brown JS, Duguid P (1991) Organizational learning and communities-of-practice: toward a unified view of working, learning and innovation. Organ Sci 2(1):40–57
5. Brown JS, Duguid P (2000) The social life of information. Harvard Business School Press, Boston
6. Bryant SL, Forte A, Bruckman A (2005) Becoming Wikipedian: transformation of participation in a collaborative online encyclopedia. In: Proceedings of the 2005 international ACM SIGGROUP conference on supporting group work, ACM, New York, p 10

7. Bui T, Cho S, Sankaran S et al (2000) A framework for designing a global information network for multinational humanitarian assistance/disaster relief. Inf Syst Front 1(4):427–442

8. Burt RS (2004) Structural holes and good ideas. Am J Sociol 110(2):349–399

9. Cataldo M, Wagstrom PA, Herbsleb JD et al (2006) Identification of coordination requirements: implications for the design of collaboration and awareness tools. In: Proceedings of the CSCW 2006, Alberta

10. Child J (1972) Organizational structure, environment and performance: the role of strategic choice. Sociology 6(1):1–22

11. Chudoba KM, Wynn E, Lu M et al (2005) How virtual are we? Measuring virtuality and understanding its impact in a global organization. J Inf Syst 15:279–306

12. Crandall D, Cosley D, Huttenlocher D et al (2008) Feedback effects between similarity and social influence in online communities. In: Proceedings of the 14th ACM SIGKDD international conference on knowledge discovery and data mining, ACM, New York, pp 160–168

13. Crowston K, Howison J (2005) The social structure of free and open source software development. First Monday, p 10

14. Crowston K, Wei K, Howison J et al (2012) Free/Libre open source software development: what we know and what we do not know. ACM Comput Surv 44(2):7

15. Crowston K, Wei K, Li Q et al (2006) Core and periphery in Free/Libre and Open Source software team communications. In: HICSS '06. Proceedings of the 39th Annual Hawaii international conference on system sciences, vol 6, Waikoloa, p 118

16. Crowston K, Wiggins A, Howison J (2010) Analyzing leadership dynamics in distributed group communication. In: HICSS '10. Proceedings of the 43rd Annual Hawaii international conference on system sciences, pp 1–10

17. Dourish P (2004) What we talk about when we talk about context. Pers Ubiquit Comput 8:19–30

18. English R, Schweik CM (2007) Identifying success and tragedy of FLOSS commons: a preliminary classification of Sourceforge.net projects. In: FLOSS '07. First international workshop on emerging trends in FLOSS research and development, Minneapolis

19. Fine GA, Harrington B (2004) Tiny publics: small groups and civil society. Sociol Theory 22(3):341–356

20. Fleming L, Waguespack DM (2007) Brokerage, boundary spanning, and leadership in open innovation communities. Organ Sci 18(2):165

21. Freeman LC (1979) Centrality in social networks conceptual clarification. Soc Networks 1(3):215–239

22. Geiger RS, Ribes D (2011) Trace ethnography: following coordination through documentary practices. In: Proceedings of the 2011 44th Hawaii international conference on system sciences, IEEE Computer Society, Washington, DC, pp 1–10. doi:10.1109/HICSS.2011.455

23. Goggins SP, Gallagher M, Laffey J (2011) Completely online group formation and development: small groups as socio-technical systems. Inform Technol People 24(2):104–133. doi:10.1108/09593841111137322

24. Goggins SP, Laffey J, Amelung C (2011) Context aware CSCL: moving toward contextualized analysis. In: CSCL 2011, Hong Kong

25. Goggins SP, Laffey J, Amelung C et al (2010) Social intelligence in completely online groups. In: IEEE international conference on social computing, Minnesota, 20–22 August 2010, pp 500–507

26. Goggins SP, Mascaro C, Valetto G (2013) Group informatics: a methodological approach and ontology for socio-technical group research. J Assoc Inform Sci Technol 64(3):516–539

27. Goggins SP, Valetto G, Mascaro C et al (2013) Creating a model of the dynamics of socio-technical groups using electronic trace data. User Model User-Adap Inter 23(4):345–379

28. Goggins S, Mascaro C, Mascaro S (2012) Relief after the 2010 Haiti Earthquake: participation and leadership in an Online Resource Coordination Network. In: Proceedings from ACM conference on computer supported cooperative work, Seattle, pp 57–66

29. Goggins S, Mascaro C, Valetto G (2012) Group informatics: a methodological approach and ontology for understanding socio-technical groups. J Assoc Inform Sci Technol 64(3):516–539
30. Goggins S, Valetto P, Mascaro C et al (2012) Creating a model of the dynamics of socio-technical groups. User Model User-Adap Interact J Pers Res 23(4):345–379
31. Gong L, Teng CY, Eecs AL et al (2011) Coevolution of network structure and content. WebSci '12. In: Proceedings of the 3rd annual ACM web science conference, Illinois, Evanston, 22–24 June 2012
32. Gross T, Stary C, Totter A (2005) User-centered awareness in computer supported cooperative work systems: structured embedding of findings from social sciences. Int J Hum Comput Interact 18(3):323–360
33. Healy PGT, White G, Eshghi A et al (2007) Communication spaces. Comput Supported Coop Work 16
34. Howison J, Wiggins A, Crowston K (2012) Validity issues in the use of social network analysis with digital trace data. J Assoc Inform Syst 12(2):767–797
35. Howley I, Mayfield E, Rosé CP (2012) Linguistic analysis methods for studying small groups. Routledge, London
36. Huffaker DA, Teng CY, Simmons MP et al (2011) Group membership and diffusion in virtual worlds. In: IEEE '11, Massachusetts, Boston, 9–11 October 2011
37. Kiesler S, Boh WF, Ren Y et al (2005) Virtual Teams and the geographically dispersed professional organization. In: Extending the contributions of Professor Rob Kling to the analysis of computerization movements California, Irvine, 11–12 March 2005
38. King G (2011) Ensuring the data-rich future of the social sciences. Science 331(6018):719–721. doi:10.1126/science.1197872
39. Kittur A, Chi EH, Pendelton BA et al (2007) Power of the few vs. wisdom of the crowd: Wikipedia and the rise of the bourgeoisie. In: CHI '07: 25th annual ACM conference on human factors in computing systems. California, San Jose, April 28-May 3 2007
40. Kobsa A (2001) Generic user modeling systems. User Model User-Adap Inter 11(1):49–63
41. Lee CP (2007) Boundary negotiating artifacts: unbinding routine of boundary objects and embracing chaos in collaborative work. Comput Supported Coop Work 16:307–339
42. Livne A, Simmons MP, Adar E et al (2011) The party is over here: structure and content in the 2010 election. In: Fifth international AAAI conference on weblogs and social media, Barcelona
43. Mascaro C, Goggins S (2011) Brewing up citizen engagement: the coffee party on Facebook. In: Proceedings of the communities and technologies, ACM, Brisbane
44. Mcculloh I (2009) Detecting changes in a dynamic social network. ProQuest, Ann Arbor
45. Mockus A, Fielding RT, Herbsleb JD (2002) Two case studies of open source software development: Apache and Mozilla. ACM Trans Softw Eng Methodol 11(3):309–346
46. Newman MEJ (2004) Analysis of weighted networks. Phys Rev E 70(5):056131
47. O'Leary M, Orlikowski W, Yates J (2002) Distributed work over the centuries: trust and control in the Hudson's Bay Company, 1670–1826. Distributed Work, 27–54
48. Olson G, Herbsleb J, Rueter H (1994) Characterizing the sequential structure of interactive behaviors through statistical and grammatical techniques. Hum Comput Interact 9(4):427–472
49. Orlikowski W (2002) Knowing in practice: enacting a collective capability in distributed organizing. Organ Sci 13:249–273
50. Palen L, Anderson K M, Mark G et al (2010) A vision for technology-mediated support for public participation & assistance in mass emergencies & disasters. In: Proceedings of the 2010 ACM-BCS visions of computer science conference, Edinburgh, 13–16 April 2010
51. Petrinja E, Sillitti A, Succi G (2010) Comparing OpenBRR, QSOS, and OMM assessment models. Open Source Software, New Horizons, pp 224–238
52. Powell A, Piccoli G, Ives B (2004) Virtual teams: a review of current literature and directions for future research. Data Base Adv Inform Syst 35(1):6–36
53. Stahl G (2006) Group cognition: computer support for building collaborative knowledge. MIT Press, Boston

54. Star SL, Griesemer JR (1989) Institutional ecology, 'translations' and boundary objects: amateurs and professionals in Berkeley's museum of vertebrate zoology, 1907–39. Soc Stud Sci 19(3):87–420
55. Strijbos JW, De Laat M, Martens RL et al (2005) Functional versus spontaneous roles during CSCL. In: CSCL 2005, pp 647–657
56. Teng CY, Gong L, Livne A et al (2011) Coevolution of network structure and content. In: Proceedings of the 3rd annual ACM web science conference, ACM, New York, pp 288–297
57. Terveen L, McDonald DW (2005) Social matching: a framework and research agenda. ACM Trans Comput Hum Interact 12(3):401–434
58. Twidale MB (2005) Over the shoulder learning: supporting brief informal learning. Comput Support Coop Work (CSCW) 14(6):505–547. doi:10.1007/s10606-005-9007-7
59. Watson-Manheim MB, Chudoba KM, Crowston K (2002) Discontinuities and continuities: a new way to understand virtual work. Inform Technol People 15(3):191–209. doi:10.1108/09593840210444746
60. Wenger E (1998) Communities of practice: learning, meaning and identity. Cambridge University Press, Cambridge
61. Wiggins A, Crowston K (2010) Reclassifying success and tragedy in FLOSS projects. Open Source Software, New Horizons, pp 294–307
62. Wu J, Goh KY (2009) Evaluating longitudinal success of open source software projects: a social network perspective. system sciences. In: HICSS '09: 42nd Hawaii international conference, Hawaii
63. Zanetti MS, Scholtes I, Tessone CJ et al (2013) Categorizing bugs with social networks: a case study on four open source software communities. In: Proceedings of the 35th international conference on software engineering, San Francisco, pp 1032–1041, 18–26 May 2013
64. Zanetti MS, Scholtes I, Tessone CJ, Schweitzer F (2013) The rise and fall of a central contributor: dynamics of social organization and performance in the GENTOO community, In: Proceedings of the international workshop on cooperative and human aspects of software engineering, San Francisco, pp 49–56
65. Zhou M, Mockus A (2012) What make long term contributors: willingness and opportunity in OSS community. In: Proceedings of the 34th ICSE. Zurich, pp 518–528, 2–9 June 2012

Chapter 9
Anonymity, Immediacy and Electoral Delegation in Socio-Technical Computer Systems

Jean Botev

Abstract Collective decision-making is a key concern for every social group; the clarity, effectiveness and participatory characteristics of the process are central to democratic societies and procedures.

Networked computer systems in general and the increasing attention to social aspects in their purpose and design offer the individual novel means for participation, but also entail specific systemic problems. These can be either variations of existing general sociological and political issues, or arising also from the system's characteristic technical design and structure. This article provides an overview of systems for collective and dynamic decision-making with their peculiarities, focusing on the three core interacting aspects: anonymity, immediacy and electoral delegation.

9.1 A Brief History of Delegated Voting

Political decision-making is a complex process, not only when individually contemplating simple binary decisions, but specifically when turning multimodal and -dimensional, i.e., dealing with a range of options in a larger group of individuals.

Since the dawn of Athenian democracy, various participatory approaches, mostly based on specific voting processes and encompassing different levels of directness, have been developed, proposed and applied.

One such way of facilitating the collective decision-making process is *delegated voting*, also known as *proxy voting*. In this voting method, the individual participants to a balloting procedure with regard to a certain question or issue can choose to delegate their vote to another participant so that a vote can be cast for them in their absence, also including the possibility of abstention or invalidity.

While the idea of delegated voting—as deployed currently—dates back to the nineteenth century [3] with its telecommunications-aided form initially described as

J. Botev (✉)
Faculty of Science, Technology and Communication, University of Luxembourg,
6, rue Richard Coudenhove-Kalergi, L-1359 Luxembourg, Luxembourg
e-mail: jean.botev@uni.lu

© Springer International Publishing Switzerland 2014
K. Zweig et al. (eds.), *Socioinformatics - The Social Impact of Interactions between Humans and IT*, Springer Proceedings in Complexity,
DOI 10.1007/978-3-319-09378-9_9

a concept already in the late 1960s [10,13], today's networked computer technology enables its realization without logistic restrictions altogether, thus increasingly blurring the concepts of direct, participatory democracy and indirect, representative democracy, as well as implicit and explicit voting.

9.2 Key Factors in Electoral Delegation

A recent experimental study reveals that endogenously implemented electoral delegation works well with preliminary communication between the individual participants, leading to overall efficient and equitable results [5].

However, it is important to note that not only the initial communication between delegator and (potential) delegatee is decisive, but in large-scale, dynamic socio-technical systems there is also a permanent feedback loop between the two, and the actual decisions themselves feed back on the constituents in the system, as well.

This low-level reflexivity is both interesting and challenging in itself, but when it occurs on a global scale networked computer system such as the Internet with potentially geographically and temporally decoupled users, it adds another layer of complexity to the overall dynamics of the process.

In addition to the actual delegation process, the two key factors *anonymity* and *immediacy* discussed in more detail in the following need to be considered in the analysis of specific socio-technical systems.

Exploring these aspects and their often circular interdependencies with regard to decisions, behaviors, communication and other processes is an essential building block both in and for the design of socio-technical computer systems, both if including explicit voting and delegation mechanisms or not.

9.2.1 Immediacy

The formation of hierarchies is an integral part of the delegation process. An important aspect for the dynamic bottom-up formation of efficient hierarchical structures is the combination of meritocratic characteristics and immediacy of the delegation, i.e., the amount of time between granting and withdrawing deciding votes for both the agenda setting and the actual decision-making, which allows for the fine-grained transition between participatory and representative modes.

9.2.2 Anonymity/Pseudonymity

In current socially aware systems, the role of anonymity and/or pseudonymity is highly controversial. Traditionally, electoral anonymity implies freedom from persecution and freedom of speech while legitimization is taken care of by an external authority which is cut out from the decision-making process as such. In engineered networked computer systems that explicitly involve social structure, however, also the authorization or legitimization becomes critical as it is a part of or at least directly connected to the system itself, causing potential for misuse both internally and externally—a dilemma that needs to be resolved in order to establish the necessary level of acceptance with regards to the perception and use of such systems and the credibility of the obtained results.

9.3   Existing Participatory Systems

Most socio-technical systems have a participatory character and therefore involve voting procedures. These are often limited to a majority or counting vote without the possibility of delegation, such as for conflict resolution in version control systems or bug tracking in collaborative software development. In Bugzilla, for instance, the voting feature provides a number of votes to users which they can allocate to bugs as they deem appropriate [2].

More complex collective decision-making involving delegation as outlined in Sect. 9.1 can be found in dedicated systems. With the proliferation of social networking and Internet use in general, the design and potential of such systems was discussed already at an early stage, such as in [4, 9]. Currently, research efforts in that area mainly focus on security aspects [14] and on the integration of social network information with participatory systems to allow for an improved efficiency of the decision-making process.

Recent proposals include the *Smartocracy* [12] project that suggests implicit delegation by combining existing social graphs with the delegation process, conceptually following the idea of so-called *mental maps* [6] in order to algorithmically form a traceable decision network. Another approach is taken for instance in [1], where the concept of *viscous democracy* is discussed. Here, information from social networks is used for calculating scores that further influence the voting power also within an ongoing delegation.

However, a series of web-based participatory systems is already successfully deployed. They are mainly used for general discussions, polling and generic voting predominantly in the context of political parties or initiatives, and, more specifically, in community management and the organizational sector. Once a submitted issue or topic reaches a critical interest level among the users of the system and thus gains the necessary momentum, when crossing this predefined threshold the users begin the voting and delegation process. At the end lies a decision that ideally will be acted upon or at least in some way influence actual policy-making.

In Germany, there are essentially two main software platforms in use for online collaborative decision-making which are maintained and supported by different non-profit organizations: *Adhocracy*[1] from the Liquid Democracy e.V. and *Liquid Feedback*[2] operated by the Interaktive Demokratie e.V./Public Software Group.

The latter platform is used for running the German Pirate Party's internal forum for debate, which at the moment constitutes the largest deployed system of that kind with more than 10,000 members and almost 19,000 initiatives or proposals structured into roughly 3,500 topics [11]. However, many other parties, initiatives and organizations covering a wide political and social spectrum run their own platforms using similar technologies.

While some recent studies dedicated to liquid feedback and the concept of liquid democracy [7, 8] also discuss aspects and mechanisms such as the transitivity of delegation and its effect on topic selection and agenda-setting, a quantitative analysis of deployed systems as well as the connections and cross-effects between the various constituents remain largely unexplored.

9.4 Perspectives

Taking into account these considerations, this article intends to highlight the importance of the way socio-technical networked computer systems and their users interact with each other and contribute to their interdisciplinary understanding.

Electoral delegation as a central mechanism to enable and facilitate the decision-making process is already well established. Still, many related aspects in dynamic systems are not fully understood yet.

An important issue to investigate in this regard is how and to what extent the increased immediacy in such participatory systems affects the users. The inherent higher dynamics, i.e., more frequent revocations and reallocations of the delegation appear to intensify the pressure on the delegatees.

Preliminary analyses also indicate that some notions might prove to be actually counterintuitive. For example, anonymity and pseudonymity in closed or restricted systems with a single account per user appear to have only a limited temporal effect, with a Hawthorne-like reactivity setting in soon after.

The understanding of the complex dependencies and interactions between these and other constituents is crucial both in and for the design of any type of socio-technical system. While there is a large body of work on traditional political and socio-economical systems, we are only beginning to connect all the dots and to master the processes underlying novel socio-technical networked computer systems and their intrinsic challenges.

---

[1]http://www.adhocracy.de

[2]http://www.liquidfeedback.org

References

1. Boldi P, Bonchi F, Castillo C, Vigna S (2011) Viscous democracy for social networks. Commun ACM 54(6):129–137
2. Bugzilla Bug Tracking System: The Bugzilla Guide – 4.4.1 Release (2014). http://www.bugzilla.org/docs/4.4/en/pdf/Bugzilla-Guide.pdf
3. Carroll L (1884) The principles of parliamentary representation. Harrison and Sons, London
4. Ford B (2002) Delegative democracy. http://www.brynosaurus.com/log/2002/0515-DelegativeDemocracy.pdf
5. Hamman JR, Weber RA, Woon J (2011) An experimental investigation of electoral delegation and the provision of public goods. Am J Polit Sci 55(4):738–752
6. Heylighen F (1999) Collective intelligence and its implementation on the web: algorithms to develop a collective mental map. Comput Math Organ Theory 5(3):253–280
7. Jabbusch S (2011) Liquid Democracy in der Piratenpartei. University of Greifswald, Germany, pp 1–186
8. Kuhn I (2013) Liquid democracy: Chancen und Grenzen einer neuen Form der Bürgerbeteiligung. European University Viadrina Frankfurt (Oder), Germany, pp 1–127
9. Leggewie C, Bieber C (2001) Interaktive Demokratie – Politische Online-Kommunikation und digitale Politikprozesse. Aus Politik und Zeitgesch (B 41–42):37–46
10. Miller JC (1969) A program for direct and proxy voting in the legislative process. Public Choice 7(1):107–113
11. Piratenpartei Deutschland (2014) LiquidFeedback in der Piratenpartei Deutschland. https://lqfb.piratenpartei.de
12. Rodriguez M, Steinbock D, Watkins J, Gershenson C, Bollen J, Grey V, DeGraf B (2007) Smartocracy: social networks for collective decision making. In: Proceedings of the 40th annual Hawaii international conference on system sciences (HICSS 2007), Waikoloa, pp 90–97
13. Tullock G (1967) Toward a mathematics of politics. University of Michigan Press, Ann Arbor
14. Zwattendorfer B, Hillebold C, Teufl P (2013) Secure and privacy-preserving proxy voting system. In: 10th international IEEE conference on e-business engineering (ICEBE 2013), Coventry, pp 472–477

Chapter 10
Towards Acceptance of Socio-technical Systems – An Emphasis on the Requirements Phase

Joerg Doerr

Abstract This paper discusses some essential reasons why socio-technical systems do not get the acceptance by end-users that is typically needed for a successful usage of the systems. One major prerequisite to achieve sufficient acceptance is to ensure a specific treatment of end-users, and stakeholders in general during the requirements engineering phase. The paper illustrates examples how requirements engineering activities for current as well as for future applications can help to increase the acceptance of socio-technical systems.

10.1   Introduction

In the development of socio-technical systems, the requirements engineering phase is essential for the later acceptance of the systems by its stakeholders. Important, especially non-technical aspects of the system such as ethical, legal, as well as cultural aspects can be identified in this phase and serve as an important requirements basis for the further system development. Furthermore, they serve as the basis for quality assurance, i.e., are the baseline to evaluate a socio-technical product before it is released to its end-users.

The discipline of Requirements Engineering as sub-discipline of Software Engineering is established in academia since about 20 years [1]. Nevertheless, we can find a multitude of socio-technical systems in our society that find no or only low acceptance at its end-users. The reasons for this are discussed in the following by taking the perspective of industry-related reasons as well as academia-related reasons.

J. Doerr (✉)
Fraunhofer IESE, Fraunhofer-Platz 1, 67663 Kaiserslautern, Germany
e-mail: joerg.doerr@iese.fraunhofer.de

© Springer International Publishing Switzerland 2014
K. Zweig et al. (eds.), *Socioinformatics - The Social Impact of Interactions between Humans and IT*, Springer Proceedings in Complexity,
DOI 10.1007/978-3-319-09378-9_10

145

Although we can see an increasing usage of requirements engineering methods, techniques and tools in the recent years, there is still a low spread of requirements engineering techniques and methods in industry. Additionally, if requirements engineering is performed in industry, it is often treated with a rather technical perspective. This means that only few approaches in industry make use of methods that have a strong emphasis on stakeholder analysis, or integrate with user centered design [2]. Rarely, a structured stakeholder analysis is used. A third reason for low acceptance is the low spread of end-user centricity in general throughout the software development lifecycle. E.g., the usage of personas [3] during all stages of software development in the offices of software engineers can help to propagate the paradigm of end-user centric development.

In academia, we can find few approaches that combine user centered design approaches with traditional requirements engineering approaches. Even though the disciplines of requirements engineering and usability engineering have many activities, principles and even produced artifacts in common, one can rarely spot scientific papers that integrate both worlds such as [4–6]. A second reason that leads to systems of low acceptance is the fact that the field of user experience is still in its infancy and needs to advance faster to provide more methodological guidance to the software engineering communities. Finally, we see a lot of socio-technical systems that basically have a large subset of members of our society as end-users, but only few of them are involved into the system development during the requirements engineering phase. Those "representative end-users" are then identified and involved in the requirements engineering during product development. All others do typically not have any possibility to influence the software development. With the advent of social media, new approaches such as the usage of crowd-sourcing in requirements engineering and enabling the end-users to provide direct requirements [7] should be researched more intensively. The remainder of this paper emphasizes the use of requirements engineering activities for current and future applications: the use of concepts like task-oriented requirements engineering to increase acceptance for current systems and an outlook on how to use crowd-sourcing in requirements engineering for future applications.

10.2 Using RE to Increase Acceptance for Current Applications

In this chapter, two essential principles for requirements engineering that can increase the acceptance of socio-technical systems are outlined: the deployment of a holistic view on requirements engineering, taking technical, business and end-user concerns into account, and the usage of task-orientation, exemplified by means of the TORE approach.

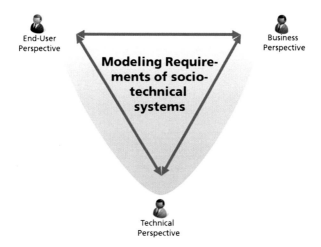

End-User
Perspective

Business
Perspective

**Modeling Require-
ments of socio-
technical
systems**

Technical
Perspective

Fig. 10.1 Essential stakeholder perspectives on socio-technical systems

10.2.1   Using a Holistic View During RE

One key challenge in modeling the requirements of socio-technical systems that
shall provide high acceptance of end-users is to take into account the different perspectives on the socio-technical system: Basically, one can distinguish
the requirements from a business, an end-user, and a technical perspective (see
Fig. 10.1). If too much emphasis is given to technical solutions, and modeling
focuses on the IT system only, important aspects of the organizational context or
the end-user requirements are often neglected. This is one major reason for the
perception of insufficient business – IT alignment and also low acceptance of the
system by end-users and business stakeholders. Therefore, requirements models of
socio-technical systems need to represent information from all three perspectives.
During elicitation and integration of this information, the necessary interdisciplinary
working of the teams is frequently perceived as a major obstacle for efficient
modeling. In practice, smooth interaction between these three perspectives based on
integrated requirements modeling is a key success factor for efficient and effective
requirements engineering that can lead to higher acceptance of the final system.

10.2.2   Using Task-Orientation in RE

About 10 years ago, the TORE (Task- and Object-oriented Requirements Engineering) approach for modeling interactive systems emerged as a result of a
systematic analysis of existing approaches for modeling interactive systems [8].
This approach was updated in recent years to cope with the challenges of modeling

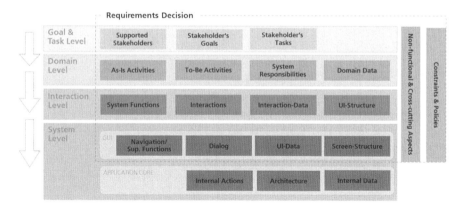

Fig. 10.2 Concepts in the TORE framework

complex socio-technical systems. The description of TORE in this paper is modified version of the description of TORE in [4]. TORE is a decision framework, which encapsulates decisions on four different levels of abstraction that typically have to be made during requirements engineering for socio-technical systems (see decision points in Fig. 10.2). The decisions correspond to requirements concepts that can be modeled for a socio-technical system. They are independent of concretely used requirements processes or notations, allowing high applicability in many different contexts. For each concept, it is typical that a requirements specification contains artifacts that model information about these concepts. The concepts of TORE will be described in detail in the following.

At the Goal & Task Level, the first decision point is *Supported Stakeholders.* Deciding which stakeholders should be supported by a system to be developed is usually one of the initial decisions to be made and an essential one to ensure that all intended stakeholders will accept the future product. Stakeholders are analyzed with regard to the previously mentioned business, technical and end-user dimensions. Notations used to model this decision are typically stakeholder maps as used in [9] (see also Fig. 10.3), stereotypical user descriptions such as personas [10], or simple role descriptions.

The second decision point is to capture which *Stakeholder's Goals* exist and shall be supported by the system. Understanding the stakeholder's goals is essential to reach high acceptance. It is the first abstraction level to justify all future interactions and system functions of a socio-technical system. The functions themselves and how they are implemented must be clearly motivated by indirect or direct traces back to the stakeholder's goals. TORE models goals of organizations (business goals) as well as goals of users (individual goals). Typical notations used for modeling goals are notations used in methods such as KAOS [11], i* [12], or simple AND/OR goal refinement trees. Typically, the functional goals are refined into S*takeholder's Tasks.* In a simple system, the *Stakeholder's Tasks* include the tasks of the users,

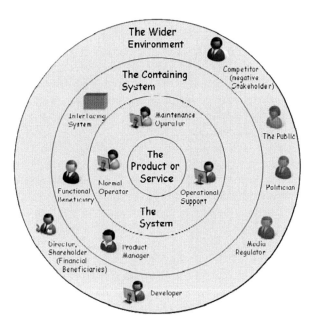

Fig. 10.3 Basic techniques like the stakeholder onion model can support acceptance during requirements engineering

while in complex business information systems, this decision point is the hierarchy of business processes.

At the Domain Level, each *Stakeholder's Task* is then refined into its *As-Is Activities*, i.e., the description of how tasks and business processes are currently performed without the system to be developed. The knowledge about the current situation makes later development stakeholders aware of where the end-users come from and what they are used to and what will be completely new system concepts. This is essential to plan for change management if essential process steps are changed. This is often neglected in many system development approaches. In contrast to the As-Is Activities, the *To-Be Activities* describe the tasks or business processes as they should be carried out when the system to be developed is in place. The typical notation used to model the *As-Is and To-Be Activities* are process modeling notations such as BPMN [13], EPCs [14], or UML Activity diagrams [15]. By modeling the *System Responsibilities,* one then determines which of the *To-Be Activities* are performed automatically, and which are performed only by humans, respectively by humans using system support. Often, the *To-Be Activities* and the *System Responsibilities* are determined at the same time. Furthermore, *Domain Data* determine which data is handled on the Domain Level, respectively within the *To-Be Activities*. Typical notations for modeling the *Domain Data* are ER Diagrams or UML class diagrams.

At the Interaction Level, the *Interactions* define for all system-supported *To-Be Activities* what the concrete usage of a system by a human should look like. This is an essential activity to reach high usability of the system. The interaction of the human with the system is often left as implicit decision to the development stakeholders. Especially if the development stakeholders are not used to the domain or do not have a clear impression of the future end-users, inappropriate decisions will be made for this decision point. So we see it as essential to make these decisions explicitly during the requirements phase. Typical notations used to model decisions of this decision point include Use Cases [16] or other scenario techniques. For all *System Functions* that are identified during the *To-Be Activities* and *Interactions*, the *System Functions* then describe the corresponding details (visible behavior, input, output, etc.). Furthermore, the *Interaction Data* determine the data used in *Interactions* and *System Functions*. Hence, they are typically a refinement of the *Domain Data*, using similar notations. With regard to early UI design, the *UI-Structure* is a first logical grouping of functions and data, but with neither a detailed layout nor a modality decision. Typical notations used to document these decisions are workspaces as proposed in [17].

As explained before, TORE already offers strong integration of usability concerns into the requirements models to increase acceptance. Nevertheless, requirements for other important qualities such as performance, maintainability, or portability need to be modeled in order to create high-quality socio-technical systems. We depicted the need for such models with the *Non-Functional & Cross-Cutting Aspects* decision point as an extension of the classical TORE framework in Fig. 10.2. Furthermore, *Constraints* for the system development as well as *Policies* such as Business Rules need to be documented. More detailed information on TORE in general can be found in [8].

10.3 Outlook: Using RE to Increase Acceptance for Future Applications

We see a class of socio-technical systems emerging that has a massive number of end-users. Examples from the private domain are social networks. With the advent of so-called smart software ecosystems, we will also see more and more information systems with several thousands of end-users emerge. Examples are information systems in smart cities or networked systems that include data from transportation systems. Potentially, all citizens are end-users of such systems. To increase the probability that these systems will reach high-acceptance, new requirements engineering techniques, especially for the evolution of such systems can be used. We foresee that for that purpose, crowd-sourcing will take place to support requirements elicitation. On the one hand, automatic data will be collected from the end-users' usage of the system (respecting privacy issues). We call this approach also mining-based requirements engineering (cf. Fig. 10.4) and the data will be (semi-)automatically processed. From that hypothetic requirements will be proposed and validated. On the other hand, end-users will be motivated to provide

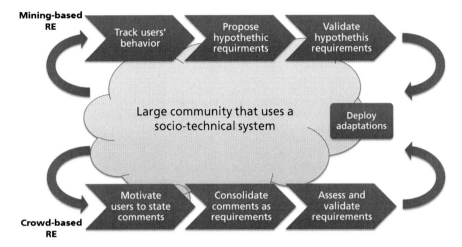

Fig. 10.4 Usage of mining-based and crowd-based RE to increase acceptance of future systems

intentional and direct feedback (e.g., via videos, pictures or textual requirements) [7] to gather requirements. Finding an efficient way to integrate this information into the modeling artifacts of our current systems for these kinds of complex socio-technical systems will be a major challenge in future research.

10.4 Summary

In this paper, we argued that the requirements engineering phase is an essential phase to achieve high end-user acceptance for socio-technical systems. But even though the discipline of RE is around for about 20 years, many systems with low acceptance exist. We argued that this is due to the fact that research, especially in the area of user experience and also the intertwining of end-user centricity and requirements engineering is still to advance. On the other hand, existing RE techniques have not made their way to industry. Based on this, this paper provided a short description of two basic principles for RE that can help to achieve higher acceptance: a holistic perspective on RE, taking into account business, end-users and technical stakeholders on the one side and the task-oriented requirements engineering paradigm with the TORE approach as example on the other side. For future application systems, we envision that crowd-sourcing can help to achieve higher acceptance. For this, mining-based RE approaches can be combined with crowd-based feedback approaches.

References

1. IEEE international symposium on requirements engineering (RE'93) (1993) San Diego, CA, USA
2. Mayhew DJ (1999) The usability engineering lifecycle: a practitioner's handbook for user interface design. Morgan Kaufmann, San Francisco. ISBN 1558605614
3. Pruitt J, Grudin J (2003) Personas: practice and theory. In: DUX '03 Proceedings of the 2003 conference on designing for user experiences, ACM, New York, pp 1–15
4. Adam S, Doerr J, Eisenbarth M, Gross A (2009) Using task-oriented requirements engineering in different domains – experiences with application in research and industry. In: Requirements engineering conference, 2009. RE '09. 17th IEEE international requirements engineering conference, Atlanta, pp 267–272
5. Lauesen S (2002) Software requirements: styles and techniques. Addison-Wesley Longman, Amsterdam
6. Doerr J, Hartkopf S, Kerkow D, Landmann D, Amthor P (2007) Built-in user satisfaction – Feature appraisal and prioritization with AMUSE, 15th IEEE international requirements engineering conference. RE 2007 – Proceedings, IEEE Computer Society, Los Alamitos, pp 101–110
7. Seyff N, Graf F, Maiden N (2010) Using mobile RE tools to give end-users their own voice. In: 18th IEEE international requirements engineering conference (RE 2010), Sydney, pp 37–46
8. Paech B, Kohler K (2004) Task-driven requirements in object-oriented development. In: Perspectives on software engineering. Kluwer Academic Publishers, Boston
9. Roberston S, Robertson J (2006) Mastering the requirements process. Addison-Wesley Professional, Reading
10. Cooper A, Reimann R, Cronin D (2007) About Face 3.0: The essentials of interaction design, Wiley, Indianapolis
11. Dardenne A, van Lamsweerde A, Fickas S (1993) Goal directed requirements acquisition. Sci Comput Program 20:3–50
12. Yu ESK (1997) Towards modeling and reasoning support for early-phase requirements engineering. In: Proceedings of the third IEEE international symposium on requirements engineering, IEEE, Annapolis
13. Business Process Modeling Notation Version 2.0, Feb 2013, http://www.omg.org/spec/BPMN/2.0/
14. Keller G, Nüttgens M, Scheer A-W (1992) Semantische Prozeßmodellierung auf der Grundlage Ereignisgesteuerter Prozeßketten (EPK). Universität des Saarlandes, Saarbrücken
15. Rumbaugh J et al (1998) The unified modeling language reference manual. Addison-Wesley, Reading
16. Cockburn A (2000) Writing effective use cases. Addison-Wesley, Boston
17. Beyer H, Holtzblatt K (1998) Contextual design: defining customer centered systems. Morgan Kaufmann Publishers, San Francisco

Chapter 11
Morals, IT-Structures, and Society

Wolfgang Lenski

Abstract Based on Neuser's conception of knowledge (Neuser, Wissen begreifen. Zur Selbstorganisation von Erfahrung, Handlung und Begriff. Springer, Heidelberg, 2013) a new characterization of morals is given which transfers the structure of knowledge to morals. This transformation is designed in a way such that the dynamic role of the inner functionalities is preserved. As a consequence a methodology evolves which explains the rise of morals in a society and at the same time identifies the factors that govern its internal dynamics. Especially, the influences of it-structures and technologies on the rise and change of morality can then be clearly understood. This approach thus prepares the conceptual grounds for questions of orientation, self-reassurance, and self-positioning of a society in view of the technological development.

11.1 Introduction

In the literature as well as in political or social disputations there are many concerns on the influence of computer science on society or on social or individual behaviour, respectively. Already in the 1970s of the last century Weizenbaum [22] hinted at the impact of it-structures on society. More recently, Grimmelmann [4, p. 1758] for instance stated with respect to the social development in a quite critical sense that "We are all regulated by software now". In pursuing this point Schneider [19, p. 290f] annotated that we all are inevitably concerned with the impact of it-structure on society and consequently appeals common responsibility for future conditions of social life. In the same line of argument Paech and Poetzsch-Heffter [14] recently emphasized the importance of it-applications for the social and organizational change. In [18] Rohde and Wulff hint at the socio-technical dual nature of it-artefacts and request that already their design should take this nature into account. Orwat et al. [13] note that software "increasingly establishes the rules of human interactions" [13, p. 628] and demand that still pending activities should be based on

W. Lenski (✉)
TU Kaiserslautern, FG Philosophy, 67653 Kaiserslautern, Germany
e-mail: lenski@sowi.uni-kl.de

© Springer International Publishing Switzerland 2014   153
K. Zweig et al. (eds.), *Socioinformatics - The Social Impact of Interactions between Humans and IT*, Springer Proceedings in Complexity,
DOI 10.1007/978-3-319-09378-9_11

a thoroughly performed exploration of the effects of such an extensive penetration of every day and professional world.

All these statements reflect sensitivities in science and society that are indeed crucial for the internal dispositions and the self-reassurance of members of a polity. Hence it is not surprising that all the statements urgently demand for a clarification of principles of self-constitution of men in the light of these sensitivities. There are thesis on the kind of their intertwining, though. "Moral knowledge is learned in social relationships." may be read, e.g., in [19, p. 289]. In view of methodological reflections Orwat et al. suggest that the sociological instrument of institutional research be used to investigate "which insights about implications and options of shaping can be obtained from this research field" [13, p. 628]. But actual scientific enterprises focusing on the kind of mutual relationships based on a well-founded scientific methodology and providing insight into the dynamics of social change are rare. On the other hand, such scientific explorations as a solid ground for subsequent disputations on consequences for societies are considered to be urgent. This paper intends to contribute to these needs.

What are structures of the internal disposition of an information society? Which methods are adequate to describe influences people are facing in the information age? What promotes the social acceptance of information technology and its derivates? What are the driving forces of its dynamics? These are the types of questions that will be addressed in the following. The very nature of these questions already indicates that the main focus will be on the interrelation between aspects of computer science, (information) technology, and society. More concretely, special emphasize will be given to the impact of computer science to social phenomena. It will turn out that philosophical reflections on the internal constitution of 'informed' societies—along with the kind of individual attitudes and dispositions that may be identified therein—will constitute the main key for a well-founded answer to all these questions.

So the problem addressed is to investigate how and in which ways computer science and society interact and influence each other. In the course of our investigations we will present concepts and structures that throw a new light on this very topic. More precisely, it is intended to identify fundamental structures and clarify basic phenomena that constitute and rule dependencies between these areas. Moreover, the very mechanisms of mutual influences and dependencies will be made explicit. As a result the kind of interaction will not only become intelligible but will also provide insight into its driving forces.

11.2 Socio-Informatics

Although an analysis of social phenomena along with the involvement of structures from computer science will play a central role, the present study is not meant to contribute to research in a field that could be called 'computational sociology'. More concretely, it is not our interest to apply computational or other methods stemming from the field of computer science to solve problems of a sociological nature. Instead, the specific area of interest is to investigate the influences of information technology and computer science methods to social phenomena. It will turn out that the most promising approach will originate from the perspective of a philosophical reflection involving aspects of social acceptance and evaluation. This approach will provide a conceptualization to describe the changes in social contexts that evolve from the application of it-structures and computer science methods.

Following a suggestion of Müller and Wahlster, we will rather adopt the notion of informatics than computer science in this context, since "Informatics is more than just computer science and incorporates the impact of services on society, where individuals interact with information technology." [10, p. 521]. Hence the intention of this paper is to contribute to *socio-informatics*—a comparatively new sub-field of informatics whose main research interest has been characterized by Rob Kling as "the interdisciplinary study of the design, uses and consequences of information technologies that takes into account their interaction with institutional and cultural contexts" [7].

But what should be the key concept and the main methodology to study this field? Müller and Wahlster—although well addressing the kernel of the problem—dampen our expectations. They continue as follows: "This however, cannot be maintained with an 'Internet of Things', where physical entities interact with computers and Big Data transforms known concepts of learning and adaption." [10, p. 521]. Hence according to [10] no solutions for the problems posed above may be expected neither from methods exploring big data nor from it-structures forming the basis for subsequent physical products. Similarly, Kling's statement is rather a program than a methodological guideline for analyzing tools.

In such a situation it is generally wise to look for the most fundamental aspect that is brought in into *socio-informatics* from the participating areas, i.e. from the side of sociology in our case. The key concept of sociology is instantiated in the most influential book that essentially contributed to establish sociology as a scientific discipline: Max Weber's *Wirtschaft und Gesellschaft. Grundriss der verstehenden Soziologie* [21]. In the very beginning of the first chapter Max Weber characterized the main concern of sociology as follows:

> 'Sociology' is a word which is used in many different senses. In the sense adopted here it means the science whose object is to interpret *the meaning of social action* and thereby give a *causal explanation* of the way in which the *action proceeds* and the *effects which it produces*. By 'action' in this definition is meant the human behavior when and to the extent that the agent or agents see it as *subjectively meaningful*: the behaviour may be either internal or external, and may consist in the agent's doing something, omitting to do something, or having something done to him. By 'social' action is meant an action in which the meaning intended by the agent or agents involves a relation to another person's behaviour and in which that relation determines the way in which the action proceeds. [20, p. 7].

The crucial concept in the characterization of the subject of sociological interest is thus *action*. But it is not only pure action; it is "subjectively meaningful action"— a person's behaviour. Hence socio-informatics is essentially about the impact of informatics or computer science (applications) on social phenomena with respect to personal—or as we will see later on—individual behaviour.

In the following we will argue that we may well end up with most valuable conceptions in the field of socio-informatics, if we enrich the *sociological* perspective—based on 'pure' actions—by a *philosophical* one. More concretely, we will complement pure actions with judgements. The main goal of this paper is then to demonstrate that by doing so we will indeed gain a new perspective on socio-informatics that may show most fruitful fine-structures of its inner constitution.

Now there are two special dimensions of judgements that play a role in our context: we could judge (possible) actions according to the categories *helpful/unhelpful* or *successful/unsuccessful*, respectively, i.e. according to epistemic categories. But we could also value actions via the distinction *good/bad* or *acceptable/refutable*, respectively, i.e. according to moralistic (or ethical) dimensions.

In this context it is highly significant that with Neuser's approach [11] a new conception of knowledge enters the scene which exactly pursues this perspective. It develops a new perspective on actions which is not of a sociological nature but follows the *helpful/unhelpful* or *successful/unsuccessful* line. Being not of a sociological nature by constitution, Neuser's approach nevertheless provides a powerful tool to analyze social dynamics as well as a description methodology for the emergence of actions in a society. Basically only coping with knowledge, Neuser's approach could well constitute the key concept for studies on the influence of informatics on *morals* in a society, too. This paper is meant to contribute to exactly this topic.

11.2.1 Mass Phenomena

Another phenomenon deserves attention in view of the aims of this paper. In many areas certain types of aggregations ("masses") have properties or show behaviour that are *not* a compilation of properties of the behaviour of the individual participants. Instead, other models making use of properties which are unique to the mass are much more effective or even solely appropriate and productive. To illustrate this kind of relationship the following list mentions a few typical cases:

- *Evacuation models*
  In evacuation scenarios the personal behaviour of individuals is totally irrelevant for the superior goal to study the movement of the group. Instead, an efficient evacuation of a great number of people in specific critical situations, e.g., in case of fire, is best be simulated by the flow of a viscous fluid. This model is known as the people-fluid analogy (cf., e.g., [8, p. 559]).
- *Thermodynamics*
  In the field of thermodynamics it is well known that warmth is based on the velocity of particles but it is not computable by procedures based on the

individual velocity along with their directions in space. As a consequence, a statistical approach is used instead of causal methodologies.

- *Mathematics*
  Even in mathematics we find a variant of a mass phenomenon: the reals cannot be constructed out of the rationals by finite constructions like, e.g., the rationals can be constructed out of the natural numbers—they are *incommensurable* with the rationals as the ancient Greek mathematics formulated the matter. In modern mathematics the reason for this incommensurability is obvious: a strong form of infinity is inevitably involved into the constructions of reals—be it via Cauchy-sequences, Dededkind cuts, or the like. Always essentially infinite sets are involved which must satisfy additional properties (cf., e.g., [17, ch. 12]).

- *Computer science*
  In computer science it is commonly agreed that existing methods to handle and analyze data are not sufficient in case of huge bulks of heterogeneous data that should be handled in short time under a coherent perspective: *big data* [6]. There is a demand for new methods (cf. also [12]) to exhibit new insights into abstract phenomena with the potential to change the living conditions of (groups of) persons dramatically.

A common feature of all these examples is provided by the fact that individual deviation has not necessarily an impact onto the whole. In the evacuation scenario an individual pursuing a different escape solution has no impact on the mass—at least as long as its behaviour does not result in what could be called a 'perturbation' inside the fluid affecting a greater number of others.

11.2.2   Morals and Ethics

Ethics is one of the basic disciplines of (practical) philosophy. Its main concern is reasoning on (systems of) values or norms, respectively, arguing for their validity or attempting to provide a justification of their normative claims. Its intention is to provide a theoretical framework of general principles resulting in an orientation in the 'world' and to judge possible actions accordingly.

In view of social issues ethics is an area of pure philosophical considerations with an impact on social problems. It reflects on social problems and provides orientation for the members of a society independent of specific social conditions. This implies that no forms of empiricism is traditionally associated with ethical considerations or even involved in discourses on the topic.[1] As a consequence, the relationship between ethics and society is one-directional; society may well address ethical problems and demand answers but their treatment is a matter of Philosophy.

---

[1]It does not withstand that sometimes examples taken from everyday's life are used to illustrate philosophical issues but they are meant just to complement the theoretical discussion.

In contrast to ethics morals is meant to describe the values as well as the forms of commonly accepted behavior in a society, a polity, or in a social (sub-) system like, e.g., a peer group. A study about these subjects may well be a matter of sociological research investigating social practices, attitudes, etc. Interestingly enough, the approach developed in this paper will contribute to questions of this kind insofar as the internal driving forces of the rise of behaviour are made transparent and hence could be subjected to respective studies.

One fundamental difference between ethics and morals is especially important in our context. Being a philosophical discipline ethical problems are examined by philosophical reflection. Insofar ethics can in no way be viewed as a mass phenomenon—it cannot arise from the study of a new perspective on a huge amount of individual behaviour. The consequence is important: it must rather be considered as an area that *influences* society than being a compilation or a mass phenomenon *arising from* individual actions of its members.

The situation is different for morals. Morals may indeed—at least under some special circumstances—be described as a mass phenomenon consisting of individual actions in a polity. On the other hand, it can hardly be compilated out of individual activities or personal stances in a cumulative or statistical sense. So there is a principal gap between the *detection* of some morals and their basis in personal behaviour. In view of this situation the concern of socio-informatics is on morals rather than on ethics. Especially, the rise of morals through attitudes along with investigations on influence factors especially stemming from it-applications may well constitute a point of interest of socio-informatics.

This very area of interest may now well be associated with Neuser's new conceptualization of knowledge. The aim of this paper is to demonstrate that the most central aspects of the conceptualization developed by Neuser—and basically intended for a different purpose—may well be applied to the relationship between morals and actual social behavior. Interestingly enough, Neuser's approach essentially relies on it-structures. This ties computer science, it-applications, morals, and theories of knowledge closely together.

11.3 Conceptualizations of Knowledge

In order to understand Neuser's fundamentally new approach it is helpful to compare it with other philosophical conceptions of 'knowledge'. The most prominent one goes back as far as to Platon and hence outlasted nearly 2,300 years of critical reflection. It remained unchanged in its essential parts and is thus still the basis for modern philosophical considerations on the topic.

11.3.1 The Classical Approach

In the final parts of the Platonic dialogue *Theaitetos* [16] the concept of knowledge receives special attention and is discussed by Socrates and Theaitetos. The first point to emphasize is that according to Platon knowledge is based on subjective belief. Likewise, no further disputation is given to the fact that this belief must be true in order to qualify as knowledge. These two specifications are immediately accepted by Socrates and his dialogue partner Theaitetos and thus are not discussed further. In addition, the two partners also immediately agree that these two characterizations still must be considered as being insufficient to capture the concept of knowledge in an adequate way. A personal belief that turned out to be true just by chance should not be considered as knowledge! For example, a pure guess that later on turned out to be true would hardly be considered as pre-existing knowledge—despite the fact that it became a true belief, i.e. does actually meet the conceptual requirements given so far. This leads to the conviction that knowledge would require a justification (cf. [16]).

The result is the so-called jtb-theory ('justified true belief' theory) of knowledge which constituted the basis for any conceptual assignation up to the twentieth century. It is worth mentioning, however, that the third condition (justification) turned out to be an extremely crucial one which hardly could be fulfilled in total and so became a topic of discussion up to nowadays.

Modern characterizations of knowledge haven't changed the overall specification much besides the fact that the view of subjects along with their gnosis of the 'world' as developed since Descartes has left its marks on the matter. So instead of *justified true belief* the characterization now reads as follows (see, e.g., [1, p. 39]):

A subject *S* knows the proposition *p* iff (i) *S* believes in *p*
(ii) *p* is true
(iii) *S*'s belief is justified.

Hence besides the subject-centered re-organization of the description the essential contents remains unchanged in principle. Insofar the modern characterizations share the main problems of the classical conception of knowledge:

- Platon need not discuss truth in this context, because he could rely on his own conception. But in order to use his description outside of the context of the Platonic dialogues a specific conception of 'truth' is required.
- Any justification must rely on some other previously accepted knowledge. However, no forms of absolute knowledge constituting the grounds for every specific knowledge can be found.

Besides these well-known problems some other characteristics of this conception deserve special attention. They are obviously presupposed by Platon as there is no further discussion about them in the *Theaitetos*.

- The belief must in the first place be something that can be true or false. In other words sharp criteria of fulfillment can be applied to its content.
- Knowledge must completely be accessible to conceptual reflection. Hence at least the contents of those beliefs that qualify as knowledge must be expressible in a suitable language representing conceptual relationships. Otherwise the question of justification would be senseless which, however, is not the case in the *Theaitetos*. So knowledge has the form of a proposition.
- The underlying belief as the carrier of knowledge is bound to a subject—an individual. It either originates from a human's mind or is acquired by a person. In case it is acquired it is then necessary to appropriate its contents, i.e. to make it one's own (cf. [9] for a further description of this process).
- Although knowledge is bound to a subject, justification refers to something beyond merely subjective dispositions. A such it must gain recognition via structures whose grounds are not solely subjective such as, e.g., forms of rationality.
- The kind of justification must be accessible to the subject, though, e.g., by being rational and realizing the rational arguments of the justification. Hence the origination of the subjective belief must in principle be an object of further inspection as well or at least be reliable in some sense that has gained general recognition (cf., e.g., the externalist or causal theories of knowledge [1, p. 45ff]).

Summarizing these points we find that knowledge in the Platonic sense is related to a human subject. It describes conceptual relationships for which precise satisfaction criteria do exist which are external to but comprehensible by the subject having the belief.

11.3.2 Neuser's Conception of Knowledge

Neuser's conception of knowledge [11] is totally different from the approaches that have emerged in history so far. The basis of his conceptualization are actions. Insofar Neuser's approach shares the basic constituent with sociology. This is at the same time the reason why his kind of conceptualization may be re-contextualized in social structures as we will see in the following.

But in contrast to the sociological perspective 'action' in the sense of Neuser is primarily subject to subsequent reflection and not simply understood as a form of behaviour to be studied from the position of an observer. Just in the contrary, the entanglement with reflexive structures is the main characteristics of Neuser's specification. In Neuser's understanding actions do not necessarily need to be actually performed; it is sufficient that in principle they *could* be performed. In addition, action is not just something that happens but must deliberately be performed by an entity. As such it meets the basic disposition of the philosophical theory of actions going back to Davidson [2].

Beyond this pure basis there are important differences between Neuser's and Davidson's perspective on 'action', though. As a causal theory Davidson's interest concentrates on explanations of actually performed actions. Moreover, whereas in Davidson's approach actions react on desires, Neuser's conception is closely tied to experiences as well—a feature which makes it also different to the conceptualization of 'concept' provided by the Peircean pragmatics [15][2] which on the other hand shares the relation to (possible) actions with Neuser's conception.

Now the perspective of reflection goes along with conceptual relationships connected with actions. The key concept in this context is *experience*. Opposed to actions, experiences are conceptually interpreted entities. But Neuser's conception of 'concept' does mean something clearly dissimilar to classical understandings. Concepts according to Neuser's approach are not 'clear' and 'distinct' like with Descartes and are not abstractions from concrete entities completely characterized by their foundational specification (the 'proprium') like with Aristotle. Concepts are also not truth-functions like with Frege and with modern logics.

Concepts in Neuser's theory consist of three inner components:

- An *explicit content* of meaning which comprises the immediate content.
- An *implicit or latent content* of meaning which annotates the history of the concept as well as its position in relation to other concepts.
- *Valuations* which are inevitably associated with each concept.

What simply happened while performing an action is turned into experience only in case it is grasped by a concept [11, p. 76]. These conceptual explanations constitute the ground structure of experience [11, p. 77]. Especially, concepts are prior to experience insofar as they incorporate functionalities that turn the outcome of actions (incidences) into experiences [11, p. 79].

As concepts are not precise, comparable consequences of actions can be interpreted through slight variations of the implicit content. This is the way in which conceptual development occurs: it is the result of a slight variation of the interpretation of previous experience applied to a new situation. It can then constitute the basis for new experiences or just being ignored when not considered as successful or otherwise promising (cf. [11, chapter 2]).

Coming back to knowledge, according to Neuser knowledge is interlocked with actions in a irreducible and indispensable way. Knowledge is the mutual determination of concepts, experiences and (possible) actions. We do know something in case we have the option to derive an action out of the known [11, p. 71]. Knowledge thus mediates between concepts, experiences, and actions insofar as knowledge makes experiences suitable for actions via their related concepts [11, p. 76]. Hence knowledge may be understood as the activation of past experiences for future actions. It necessarily involves concepts as they incorporate the functionality to turn actions into experience. The following figure illustrates this situation (Fig. 11.1):

---

[2]Cf. especially the so-called *pragmatic maxim* in [15, p. 293].

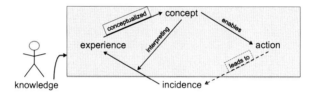

Fig. 11.1 Neuser's conception of knowledge

11.3.2.1 Individuals and Common Knowledge

So far the theory only describes *individual* actions and their relation to *individual* knowledge, concepts, and experiences. But there are more dimensions involved: communication and social mediation via shared experiences. Individuals primarily share their *experiences* with others [11, p. 80]; they communicate about their experiences and not at first sight about their concepts. They thus constitute a community of shared experiences rather than a communicating community.

This sharing of experiences give rise to a super-structure of knowledge called *common knowledge* [11, p. 95ff]. The inner constitution of this super-structure of knowledge—the *common knowledge*—reflects in principle the inner structure of individual knowledge.[3]

But the relation between (individual) knowledge and common knowledge is twofold: An individual participates in the common knowledge and communicates its own experiences which might then give rise to similar experiences by other members of the community. This sort of influence, however, is not predictable. It essentially depends on overall dispositions of the community. In times when most individual experiences are more or less successful und in fact yield the desired or expected outcome the probability of infections of more tremendous variations is rather neglectible. But in more critical times when a huge amount of members of the community are seeking for new experiences—new solutions for urgent problems—new (successful) experiences more easily find their way into the common knowledge such that others get informed and subsequently try to turn it into their own experiences.[4] It is essentially the same process that is meant when talking about the so-called 'killer applications' in technology.

Hence the relation between (individual) knowledge and common knowledge is best described as a mass phenomenon of the type discussed above, i.e. one that cannot completely be explained on the basis of the respective individual constituents. At one hand individual actions contribute to the common knowledge but the contents of the common knowledge does not in a functional sense depend on the properties of the conceptually relevant individual actions. Especially, individual variations do not

---

[3]There are slight differences that are not important, though. For example, *incidences* (the 'results' of actions) do not play a role in common knowledge.

[4]Cf. especially [11, chapter 2] for a more detailed description of this process.

immediately effect the common knowledge. Changes in the common knowledge must rather be seen as an evolving process of local variations which may or may not 'infect' the common knowledge, i.e. result in a wave of changes eventually inspiring a certain amount of participants of the community to respective new experiences. In addition, change depends on several independent factors such as, e.g., the internal disposition of the community towards the willingness to changes, the internal communication structure, the way individuals actually propagate their ideas, their authority and social status in the community, the propagation reach of stimulations, etc. So change can most probably rather be described than actually predicted. Common knowledge thus has its own dynamics which is different from the development of individual knowledge. Figure 11.2 describes this relationship:

Now summarizing the points mentioned so far it turns out that all the deficiencies of the classical approach are surmounted by Neuser's approach. At first, Neuser's conception is totally independent of classical theories of knowledge. Especially, the crucial condition of the other approaches—*justification*—is no longer present in this new context; it is simply not an issue of Neuser's theory and thus rendered obsolete. Finally, knowledge is no longer propositional.

11.3.2.2   Influences of Computer Science, Technologies, and Views of the World

Besides individual sharing of experiences, concepts, and (possible) actions there are other sources of influences to common knowledge, though. With respect to these sources of influence even the acting individuals only play a passive role insofar this kind of influence is not an outcome of any form of individual actions at all. So two totally different types of influence structures on the common knowledge must be kept separate: one stemming from conceptualized experienced individual actions and another one stemming from some external sources.

The latter ones—sources of influence totally independent from individual actions—must again be subdivided into two different influence spheres. The first

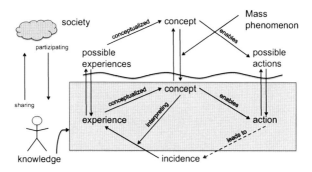

Fig. 11.2 Neuser's conception of common knowledge

one originates from the overall self-assurance of the position of men in relation to the 'outside world' (which itself must be seen as the result of a reflection on 'what there is') together with the way he experiences this relation. It is the question how people themselves understand their bare being in the 'world' they experience. This self-assurance restricts the way which 'moves' can be made at all in the actions game, i.e. the possible estimated reach and sense of one's own actions.

In the Middle Ages, e.g., truth was in its totality only recognizable by God—knowledge in its essence was principally inaccessible to men. In finding truth the task then was to understand God's wisdom as good as even possible—despite the fact that men never could actually succeed, because the principles lying behind were hidden and not accessible to men. Hence to gain true knowledge in the sense of the Platonic conception was considered as impossible in general and hence attainable only in parts and in principally restricted form.[5]

In the modern era since Descartes this kind of self-assurance has totally changed. The subject itself has entered into the center of epistemic experiences of the 'world'. As a consequence, truth, for example, is no longer only recognized and recognizable by God. Instead, the justifications as part of the knowledge conception must be realizable and judgeable by the subjects themselves. They are the new instances for which validity must be shown. As a result of this self-repositioning new forms of actions get into the reach. These are initially tested by a few at first followed by a continuously increasing number of people until they become common experience. This is the first way in which general presuppositions influence the common knowledge.

The second kind of influence stems from a totally different source. The respective line admittedly also goes back to views of the 'world'. But this time it is not the way in which subjects understand the 'world' and their own position with regard to it—it is not the epistemic view. Instead, it is the 'world' in the way it is captured by the sciences as the basis for subsequent methodological treatments. The most prominent role in view of possible impacts on society without any doubt play computer science and information technology structures. So these are the areas that deserve special attention in the following.

In every scientific fields and especially in computer science and informatics, however, it is not the 'world' in its unrestricted totality that is actually captured by the respective science. Instead, the intended and envisaged application along with theoretical interests and possibilities as well as principles of design pre-structure the treatable aspects of the whole. Accordingly, the whole variety of the being is reduced to those aspects that are in the focus of interest and that at the same time can be handled in a technological and methodological sense. 'What there is' is beforehand specified by (pre-structured) ontologies and what is treatable depends on the initial structures and design principles used to grasp certain aspects of interest

---

[5] As a side-effect this self-conception gave much authority to the institution that was considered (and widely accepted) as the legitimate authority to proclaim God's wisdom and his unbounded and limitless knowledge.

out of the totality of the unrestricted being. So it is not the totality of the 'world' but a restricted, pre-structures *view* of the 'world' that constitutes the basis for subsequent it-based functionalities. But those functionalities are the ones that finally contribute to and at the same time pre-determine our actions, experiences, and conceptualizations, i.e. our knowledge.

Again, this can happen in two ways: either in form of a direct impact on possible actions or mediated via technologies. In any case they span a space of possible actions which most probably will finally also be used. If we now want to make use of possibilities offered by computer science applications, our possible actions remain bound to the underlying design principles and information technology structures. For example, we cannot use a structured database for a request that does not meet (and cannot be derived from) its internal organization.

On the other hand we get finally acquainted to the possibilities offered by the respective structures and find some of them useful such that we start to use them quite frequently. In other words, we appropriate the given possibilities[6]—which are in turn based on the initial structures the being has been equipped with—and transform them into actions. Those actions in turn will then lead to new experiences (for example some facilitations of life) along with new concepts. Now propagating and communicating these new experiences together with the concepts will motivate others to facilitate their life in a similar way. So we may then 'infect' others, i.e. motivate them to share our experiences to the benefit that had been promised and associated with it.

Similar phenomena happen for the technologies that derive from it-structures in a more or less hidden way. They enable activities or facilitate actions that otherwise would be quite hard ones or even impossible. It's the form of propagation that finally results in new actions via enabling technologies. But exactly this way they constitute an impact on society (Fig. 11.3).[7]

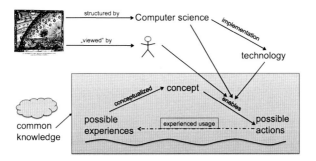

Fig. 11.3 Influences on common knowledge

---

[6]For this concept of appropriation cf. [9].

[7]Figure 11.3 includes [3].

We may now summarize as follows: it-structures and derived technologies predetermine the way in which we orientate in the 'world'. They admit or enable actions we finally find at least useful and won't miss in the future. However, those it-structures stem from a specific perspective onto the 'world'. Once such a perspectivation has penetrated the social behaviour it shows persistence and cannot easily be overcome.[8] Moreover, the entities that are considered to exist therein are also just the result of an initial design decision. So the most important impact on society is given by the way in which the 'world' is *represented* in the structures of computer science and in information technologies.

An Example

The following example on illness may illustrate this situation: Common knowledge includes conceptual descriptions of diseases which in turn include hints for suitable treatments. This is, e.g., taught during medical training at the universities. The basis for such treatments are successful healings—besides others—performed and documented by clinical studies. All this together constitutes common knowledge inside a society.

If now a person feels ill, it may consult a doctor. After identifying the disease the doctor prescribes a treatment, e.g., some pills to be taken, according to the canonical treatments associated with the concept (the disease the person is suffering from) which hopefully will eventually show the desired result.

But there are other influences. The *world wide web* contains a huge collection of more or less reliable descriptions of diseases along with suggested cures as well as personal experiences associated with illnesses, cures, and medications. From this source the person may already deduce hints on the specific kind of disease he/she is suffering from along with experiences with different kinds of treatments (but he/she better does not exclusively rely on these). If at the first glance the illness is not tremendously severe, instead of consulting a doctor the person probably chooses among the suggested treatments (and hopefully gets rid of his ailments).

In addition, there exist huge databases on medical care. All (known) diseases are catalogued and described according to classification systems such as the *International Statistical Classification of Diseases and Related Health Problems* 2010 (ICD10) (see, e.g., [5]). Software engineers have designed an appropriate database structure and functionalities to operate on these. So finally every disease is captured, catalogued and related to standard medical treatments. Other medical systems such as computer tomographs are designed to detect diseases that otherwise hardly can be detected at all and thus have extended the range of possible actions, i.e. *treatments* in this case. All these structures capture aspects and phenomena that are considered central, represent elementary properties of diseases, treatments, evaluations, etc. in suitably organized information systems and finally organize

---

[8]It is more probable that it will be superseded by a subsequent technological application.

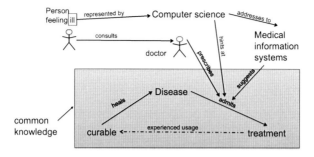

Fig. 11.4 An example of medical care

all this staff in a technical and administrative way. *Imaging methods* are ready to support the doctors in their diagnosis as well as in their attempt to detect the most promising and successful treatment (which in turn may also be performed by suitable technologies).

Computer (science) driven systems thus have a tremendous influence on the health system of a society. Medical subsystems either suggest treatments to the doctors or do even apply the suggested treatments by themselves. So the doctors need not only rely on their personal experiences and education but also get support from structures mediated by computer science along with respective technologies (Fig. 11.4).

11.4 A New Theory of Morals

A new theory of morals is now easily gained on the basis of the new conceptualization of knowledge due to Neuser. Only a few parameters have to be adjusted accordingly. The basis, however, remains still the same: actions. But besides exposing the outcome of actions—their incidences—to (epistemic) reflection[9] moralistic categories such as *good/bad* or *acceptable/refutable* may be applied with equal right. So in parallel with epistemic experiences the individuals make moralistic experiences. This is just another dimension associated with (the result of) actions which does not conflict with the epistemic view. Instead, this two perspectives may well live together giving rise to just two different perspectives onto the same entities (the incidences being the result of actions).[10]

Instead of contributing to a range of options how to act *successfully* in the 'world',[11] the individual exposes the results of his actions to moralistic categories,

---

[9]Experiences are conceptually interpreted entities.

[10]Cf. also Sect. 11.2 above.

[11]Cf. Neuser's conception of 'concept'.

i.e. how to act responsibly in the social context. Furthermore, instead of contributing to concepts ("conceptualizing") such moralistic experiences contribute to what could be called (individual) norms, i.e. the totality of experience an individual considers as acceptable. On the other hand it is this inventory of accepted experiences that leads to the judgement of (the outcome of) actions—whether they are considered as *good* or *bad*, *acceptable* or *refutable*. So Neuser's knowledge theory admits a re-contextualization according to another one of the basic dimensions of human evaluations (cf. also Sect. 11.2):

- According to the dimensions *true/false* or *successful/unsuccessful*
- According to the dimensions *good/bad* or *acceptable/refutable*.

Such an individual norm may be understood in parallel to a 'concept' in Neuser's theory. Especially, it can be viewed as bearing the following inner constitution (which is an exact structural copy of the conceptualization of 'concept'):

- An *explicit content* of evaluation
- An *implicit or latent content* which relates to the history of the norms as well
- *Valuations* (being, however, of a different kind namely according to the dimensions of *good/bad* or *acceptable/refutable* instead of *helpful/unhelpful* or *successful/unsuccessful*).

In parallel to Neuser's theory of knowledge we may also state a 'normative drift' as the result of slight variations performed by different people or by one and the same person at different times. It is just a slight local change that—in case it can be bundled with a whole variety of such local changes—eventually may well result in a rather global behaviour of members of a society (Fig. 11.5).

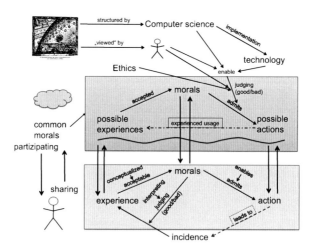

Fig. 11.5 Morals

It is then not surprising that the *common morals* arises by just the same mechanisms as in Neuser's conceptualization of knowledge. No changes are actually necessary at this point. We may just apply the same internal mechanism that turns *individual knowledge* into *common knowledge*. Because of the structural coincidence between these fields, the same driving forces can be applied to this area—practically without any change. So the inner constitution of Neuser's theory of knowledge yields a theory of morals as well. Hence it has a certain universal nature beyond the original specification.

The result may be viewed as a theory of morals instead of a theory of knowledge. However, the influence of it-applications and representation structures developed in computer science to morals is basically the same as in the field of knowledge.

11.5 Conclusion

It has been demonstrated that a structural re-interpretation of Neuser's conceptualization of knowledge yields a new characterization of (individual and common) morals. This re-interpretation still relies on the basic structures of Neuser's theory. Hence it essentially involves social phenomena, e.g., in form of infections of and communication about types of actions. As this theory of morality is influenced by it-applications and structures designed by computer science it is also a contribution to the still evolving field of socio-informatics.

As a consequence of the present explication we can also show that the demands cited at the beginning indeed are met. The theory of morals based on Neuser's theory of knowledge demonstrated that it-artefacts indeed inherit a socio-technical dual nature. Their influence on society can be viewed as a multiple paralleled appropriation of enabled actions whose carriers are the it-artefacts. The importance of it-applications for the social change is explained via the impact of it-applications on new forms of actions which result in a variation of the internal structure first on individual and subsequently on common knowledge. By transforming the whole approach to morals common responsibility is related to common morals (which may well be combined with requirements of an ethical nature). Especially, the explication given above demonstrates the penetration of every-day and professional change.

Finally, this paper demonstrates that beyond a mere conceptualization of knowledge Neuser has at the same time proposed a mechanism that is applicable to different contexts as well. The only conditions to make this mechanism work are very elementary ones: any re-interpretation of the internal constitution and the driving forces of the dynamics must be grounded on actions and judgements or evaluations, respectively. 'Concept' may then be seen as the embodiment of these judgements/evaluations equipped with two forms of connectedness to incidences as the result of actions: judging/evaluating them and at the same time being the result of such judgements/evaluations thus constituting a 'mutual determination of such embodiments, experiences and (possible) actions'.[12]

If we even dispense of forms of judgements or evaluations, we end up with just noting the bare occurrence of performed action, i.e. without further qualification. This way the given model incorporates the basic constituent of sociology.[13] So it may well be the case that Neuser's approach turns out to be fruitful in sociology as well. But this must then be the subject of a different investigation.

References

1. Baumann P (2002) Erkenntnistheorie. Metzler, Stuttgart
2. Davidson D (1963) Actions, reasons, and causes. J Philos 60:685–700
3. Flammarion C (1888) L'Atmosphere: Météorologie Populaire. Paris. Woodcut with caption: "Un missionaire du moyen âge raconte qu'il avait trouvé le point où le ciel et la Terre se touchent...", p 163. Reproduction graphics taken from http://commons.wikimedia.org/wiki/File:Flammarion_with_caption.jpg
4. Grimmelmann J (2005) Regulation by software. Yale Law J 114:1721 1758
5. International Statistical Classification of Diseases and Related Health Problems (ICD-10) version 2013. Edited by the World Health Organization. Cf. http://www.dimdi.de/static/de/klassi/icd-10-gm/
6. Klein D, Phuoc T-G, Hartmann M (2013) Big data. Inform Spektrum 36(6):319–323
7. Kling R (2007) What is social informatics and why does it matter? Inf Soc Int J 23(4):205–220
8. Korhonen T, Hostikka S, Keski-Rahkonen O (2005) A proposal for the goals and new techniques of modelling pedestrian evacuation in fires. Fire Saf Sci 8:557–567. IAFSS Symposium
9. Lenski W (2010) Information: a conceptual investigation. Information 1(2):74–118. Available at http://www.mdpi.com/2078-2489/1/2/74/pdf
10. Müller G, Wahlster W (2013) Placing humans in the feedback loop of social infrastructures. Inform Spektrum 36(6):520–529
11. Neuser W (2013) Wissen begreifen. Zur Selbstorganisation von Erfahrung, Handlung und Begriff. Springer, Heidelberg
12. Obama administration unveils "Big Data" initiative: announces $200 million in new R&D investments. Press Release of The White House, March 29 (2012) Available at www.whitehouse.gov/sites/default/files/microsites/ostp/big_data_press_release_final_2.pdf
13. Orwat C, Raabe O, Buchmann E et al (2010) Software als Institution und ihre Gestaltbarkeit. Inform Spektrum 33(6):626–633
14. Paech B, Poetzsch-Heffter A (2013) Informatik und Gesellschaft: Ansätze zur Verbesserung einer schwierigen Beziehung. Inform Spektrum 36(3):242–250
15. Peirce CS (1878) How to make our ideas clear. Pop Sci Mon 12:286–302. Reprinted in: Hartshorne C, Weiss, P (eds) (1934) Collected papers of Charles Sanders Peirce, vol. 5, paragraphs 388–410
16. Platon (1578) Theaitetos. In: Stephanus H (ed) Platonis opera quae extant omnia, 3 vols, vol 1. Paris. German translation in: Schleiermacher F (ed) (1805) Sämtliche Werke, vol. 2, In der Realschulbuchhandlung, Berlin 169–322

---

[12]Cf. the characterization of Neuser's approach given above.

[13]Cf. Max Weber's foundational remarks [20, p. 7].

17. Reiss K, Schmieder G (2007) Basiswissen Zahlentheorie. Eine Einführung in Zahlen und Zahlbereiche. Springer, Heidelberg
18. Rohde M, Wulff V (2011) Sozio-Informatik. Inform Spektrum 34(2):210–213
19. Schneider C (2013) "Das muss man immer für sich selber abwägen" oder: Das moralische Wissen von Studierenden der Informatik. Inform Spektrum 36(3):287–292
20. Weber M (1991) The nature of social action. In: Runciman WG (ed) Weber: Selections in translation. Cambridge University Press, Cambridge. German original in [21]
21. Weber M (1922) Wirtschaft und Gesellschaft. Grundriss der verstehenden Soziologie. Tübingen
22. Weizenbaum J (1976) Computer power and human reason. From judgement to calculation. W. H. Freeman, San Francisco